I0161084

DREDGING THE CHOPTANK

DREDGING THE CHOPTANK

Maryland Ghost Stories

KIMBERLEY LYNNE

Baltimore, Maryland
www.apprenticehouse.com

Copyright © 2010 by Kimberley Lynne

Library of Congress Cataloging-in-Publication Data

Lynne, Kimberley, 1961-
Dredging the choptank : Maryland's ghost stories / [compiled by] Kimberley
Lynne. -- 1st ed.
p. cm.
ISBN 978-1-934074-15-2
1. Ghosts--Maryland. 2. Haunted houses--Maryland. 3. Ghost stories,
American--Maryland. I. Title.

BF1472.U6L96 2008
133.109752--dc22

2007051091

All rights reserved. No part of this book may be reproduced or transmitted
in any form or by any means, electronic or mechanical, including photocopy,
recording, or any information storage and retrieval system, without prior
permission from the publisher (except by reviewers who may quote brief
passages).

Printed in the United States of America

First Edition

Published by Apprentice House
The Future of Publishing...Today!

Apprentice House
Communication Department
Loyola University in Maryland
4501 N. Charles Street
Baltimore, MD 21210

410.617.5265
410.617.5040 (fax)
www.ApprenticeHouse.com
info@ApprenticeHouse.com

For my grandmother,
Marjorie Louise Johnson Schuck

Acknowledgements

Thanks to Michael Angelella, Deborah Donohue Amos, Megan Anderson, Kathleen Barber, Mark Blackmon, Tish Brown, Terri Ann Ciofalo, Norrie Epstein, Neal Fandek, Lou Gieszl, David Hunt, Joe Leatherman, Betty Ann Leesburg Lang, Donald Lynne, Jeffrey Lynne, Susan Lynne, Lisa Mion, Todd Mion, Noel Schively, and Joan Weber for their guidance.

Thanks to John Benoit, Raine Bode, Shawn Brown, Andrew Ciofalo, Denise Cumor, Sid Curl, Cassandra Davis, Korinne Spence D'Amore, David Flury, Adrienne Cassara Gieszl, Thomas Hoen, Peggy Miller, Rebecca Monroe, David Orem, Shannon Parks, John Raley, B. Thomas Rinaldi, Anthony Scimonelli, Elaine Sfondias, Mr. Travis, and Dana Whipkey for their extraordinary stories.

Thanks to Margot Adler, Karen Anderson, Joseph Campbell, Joan Didon, Edith Hamilton, Mark Harp, Arlene Hirschfelder, Bertha Johnson, Paulette Molin, Alice Ann Parker, Anthony Reda, Marjorie Schuck and Christopher Weeks for expanding my mind.

Thanks to Blackwater National Wildlife Refuge, Cambridge Library, Dorchester County Arts Center, and The Eastern Shore Tribal Council for protecting Dorchester County's precious resources.

Thanks to Lillian Jackson Braun, George Carey, Helen Chappell, Thomas A. Flowers, Trish Gallagher, Vernon O. Griffen, Dickson Preston, Brice Stump and Thames Toy Store for telling Maryland folklore and expanding my heart.

TABLE OF CONTENTS

Haunted Hunting 1
Walking With Ghosts 5
A Chilled Wind 25
Push Back 43
Read Folklore 51
Fictionalized Reality 61
Black Shapes 79
Show Me the Way 87
Ghosts or No Ghosts 99
Porch Puddles 147
Bad Luck Blue Boats 165
Worm Windows 185
Human Here 205
The What 211
Fork-lore 233
Pasta Paper 237
The Devil Made Me Do It 249
I Heard No Water 269
Cancelled Czech 279
Storm Coming 283
A Human Head 293
It Floods A Lot Here 299
We Need a Carcass 331
On My Way There 335
Burial Records 347
We Believe 373

After April 2nd

HAUNTED HUNTING

Once, in a place that seems outside time, I wrote a
ghost walking tour for a small town on the Eastern Shore
of Maryland. As I collected local folklore, a Cambridge
resident named Mr. Travis told me this ghost story that
happened in his hunting lodge. Hunting's popular on
the Eastern Shore; it's rural enough for its populace to
still use weapons to catch dinner.

"It wasn't much," Travis said of the lodge. "Just a
couple of bedrooms and a kitchen and a bathroom. Up
on stilts because of all the flooding." The house stood
on its stilts on isolated Aisquith Island in the haunted,
southern underbelly of Dorchester County. Aisquith
hovers only a few feet above sea level, floating between
miles of wet fen and the Honga River. Before the lodge
was built, its low woodlands were the sacred ground of a
Native American Indian graveyard. Before John Smith
showed up in 1608, indigenous people had developed
a millennia of civilization, and, in the history of this

country, live conquering people plow dead people under.

Travis says he regularly hears children laughing when there's nothing but cattails and marsh holes for miles, and every time he returns to the lodge, the salt shaker has inexplicably spilled over. Things happen there.

One of Travis' friends stayed with his young son in the lodge. The son got up in the middle of night to get a glass of water in the kitchen. The mattress spring squeaked, and an owl hooted outside. In the living room, a strange man rocked in the rocker. He wore a plaid shirt and blue jeans and had a black plait of braided hair. He was strange only because the boy didn't know him. He thought perhaps the man was one of his father's hunting buddies; the ways of the adult world were still a mystery to the boy.

"Hello," said the boy. The rocker creaked. The man seemed to have shape and weight, like a living man.

The man nodded, and when the boy returned from the kitchen, the man was gone.

The boy tapped his sleeping father. "Where did the man go, Dad?" He asked. "The man in the living room. He looked so real."

They searched the lodge house and found no one. They looked outside into the wavering, dark pitch of the Eastern Shore night. One lone green ball of light glided over the undulating marsh grass and then vanished. The son asked to leave.

There's no sanctuary from the past; not even our living rooms are safe. I'm scared to look into my Baltimore living room late at night for fear of seeing even briefly into another dimension. My friend Korinne once slept on my couch and awoke to see a man seated in my arts and craft era sliding rocker.

"Didn't that freak you out?" I asked the next morning, aghast.

"No," she said, smiling and sipping coffee. "He seemed very happy to be here."

"What'd he look like?" I stuttered.

"Oh, I don't know," she said calmly. "Older guy, white, I think he was wearing a dinner jacket."

I realized that my ghost stories and Travis' ghost stories match the collective archetype, that they're not singular but fall into the same pattern as the rest of humanity.

Korinne was born in Detroit and lives in New Hampshire. She has the long blonde hair of a mermaid and the terrible gift of prescience. She says that something big is going to happen to me, but something happens to everyone.

Before April 2nd

WALKING WITH GHOSTS

Stories need context, geographic and historic references needed for understanding localized archetype. These ghost stories and I emerged from the same lush cradle that is the state of Maryland. I was born just south of the Mason-Dixon Line at the beginning of the Kennedy administration. I was a Junior Oriole. I was a Girl Scout. I was raised Episcopalian. I didn't consider myself Southern or Northern until I lived in Texas.

Northern states clump the state of Maryland with the South, but the Southern states clump Maryland with the North. When I lived in Dallas, I tried to explain the Mason-Dixon Line, the geographic delineation between southern Pennsylvania and northern Maryland. The boundary line was surveyed in the 1760s to settle a land dispute between two families. In 1820, Congress used the Mason-Dixon Line to separate the slave-owning states from the free states as part of the Missouri Compromise, and during the Civil War the Mason-

Dixon Line evolved into a representation of the foggy border between North and South.

"I was born south of it," I explained to blank faces. "In Baltimore. In the South."

"That's not the South," they'd drawl, one arched eyebrow up. "This is the South. Baltimore? Isn't that in one of those Yankee states? Near New York City?"

The biggest battle of the Civil War was barely ten miles from the Mason-Dixon Line. I hiked up Gettysburg's Little Round Top when I was short enough to need a leg up onto the cannons. My father, my brothers and I would strike out across the hot graveyard fields, thick with goldenrod and fat flies and steeped with the blood of 50,000 ghosts. My father was born in Fargo, North Dakota, and he taught us that the horse's legs on the battlefield statues chronicled the rider's destiny. All legs on the ground meant he survived unscathed. One leg up meant he was wounded but survived. Two legs up meant the rider bought the farm. Their battle death plot became a crucible of mud, blood and bone, the chemistry of history. Traces are still woven through the earth and our communal memory. The dead men on rearing horses fascinated me; they had a glorious crossing under the roar of battle. Their stories were told in bronze. If I stood stock still in front of them and concentrated and opened myself up to the wondrous possibility of them, I could hear the bray of stallions and the hiss of buckshot. It seemed far away and oddly beyond the next ridge at the same time. I had no idea how close it was.

Not until I lived in Texas in the late 1980s, did I realize that the War Between the States, or that altercation sometimes referred to as The War of Northern Aggression or The Civil War, was not yet over and that I was considered a Yankee, one who hailed from the Aggression side. I've always been Union philosophically, but Yankees seemed somehow harder than the people where I lived, not tougher at the core but harder on the surface. Southerners seemed fiercer at the center and softer on the outside.

I was raised in a Civil War border state in a city that has been a battlefield several times. In 1861, Maryland was a slave-owning state. Her governor Thomas Holliday Hicks owned slaves and was a native of its Eastern Shore Dorchester County. In April 1861, Confederate sympathizers attacked Union troops in Baltimore, and President Lincoln placed Maryland under martial law with cannons on Federal Hill and troops on Pratt Street. Hicks declared Maryland's Union loyalty, but Maryland's wealthy Eastern Shore was decidedly Confederate and smuggled goods and gold to Richmond throughout the Civil War. Maryland was filially split, cousin against cousin and brother against brother.

Our dead brothers still follow us. Ghost stories born during the Civil War still haunt the Eastern Shore.

Not only split by war, Maryland is geographically torn in half by the world's largest fresh water estuary, the Chesapeake Bay, dividing the state into its Eastern and Western Shores. Despite the Chesapeake Bay Bridge spanning its watery gap, Maryland's halves are forever different socially, economically and politically: Yankee vs. Confederate and industrial vs. rural. We keep to ourselves and are wary of each other.

Most Baltimorean experience of the Eastern Shore is transitory; we vacation in Ocean City and drive through the Shore to reach the Atlantic. We rarely leave Route 50 as we travel on it through the Shore; we stop to buy corn, strawberries, tomatoes, and gasoline. On those quick stops at filling stations and fruit stands, in my home state, I've felt the same itchy alienation that I've felt hovering uncomfortably in a Mississippi grocery store. I don't belong. I'm the outsider.

Someone in the back of the store whispers, "You ain't from round here, are ya?" Their sharp stares singe my clothes. Their voices and vowels are rounder. Their sky seems bigger and their air thicker. I'm amazed that we use the same currency and root for the same baseball team.

How could we possibly belong to the same state?

Separated from half my state or not, I like being a Marylander; I feel for my state the same sort of pride I hold for my country, maybe more. Certainly, I defend Maryland with a greater vengeance than America. I personally relate to a greater percentage of Marylanders than Americans. I exhibit unconscious, state-specific, jingoistic behavior. I like Delaware and Pennsylvania, but I couldn't live there. My college buddy lives in Virginia, and I wonder how he can spend all those nights there. Still, no matter how crowded and conservative Virginia seems, my allegiance to Maryland is far from logical. Maryland's a lovely place, but there's no rational reason why I should feel so strongly about her, except that she's my home, she's haunted and she shares my first name.

I am xenophobic of Virginia. Xenophobia, the fear or hatred of foreigners or of anything foreign or strange, riddles America, but strangely, Americans have few words to describe state-specific, xenophobic nationalism except parochialism with its negative connotations. We are still swayed by Darwin's strict niche rule of evolution. We like to group ourselves into same-type bunches and identify ourselves that way. We categorize and compartmentalize; it's an easier process. We like to live with people like us; it's safer. That safety is the cradle of xenophobia.

I was taught xenophobia early. My high school, Towson Senior High, played Dulaney Valley High in football, so we Towson kids were supposed to hate the Dulaney kids. It made little sense to me; I went to elementary school with a third of them. They were friends of mine. Why loathe them because they lived on the other side of Seminary Avenue? There isn't enough geographic change between both sides of Seminary Avenue to warrant the fear of difference between the inhabitants; it's a two lane road that runs through the middle of middle-class suburbia. But, certainly, there's literal gulf enough between the Maryland Western and Eastern Shores to brew that xenophobic disparity, a crevasse-sized gap between mountain foothills and sea-level marsh.

For me, loving my split, soggy state to an illogical extreme is the same as hating everything and everyone outside of it. Maybe words are constrictive boxes, and maybe none are big enough to hold the wide loyalty that people feel for home and its land, its stories and its ghosts.

When I told Chief Winter Fox of the Eastern Shore Tribal Council that I was searching for a word to define state-specific, xenophobic behavior, he understood. "For the Native American Indians," he said, "xenophobia is a matter of survival." They have to stick together to maintain their culture, and they use their stories, their stories that come from rich marsh, to maintain that culture. Geography slices our thoughts, carves our voice and shapes our hearts. Despite technology and the transitory nature of our living patterns, Californians are different than Floridians; northern Californians are different than southern Californians; Texans are different from everybody; and Western

Shore Marylanders are different from the natives of the
Eastern Shore. I recently received an email that listed
humorous definitions of state-specific behavior. *"You
live in Mississippi when you can rent a movie and buy bait in the
same store and after five years you still hear, 'You ain't from 'round
here, are ya?'"* Other people from other geographies talk
differently, and differences define our stories.

Many an August, my parents threw the three of us
kids into the back of a series of Volkswagen buses and
hung under the limitless sky of the U.S. Highway system,
visiting relatives and gaping at incredible geographic
wonders. My relatives speak a variety of dialects; they're
all over the map. My imitations of the Minnesotan
branch drove my brothers crazy.

"Tell her to stop talking like that," my older brother
would gripe, weeks after we had returned to the Eastern
Seaboard.

I missed the sound of my father's youth. I liked
wrapping my mouth around the middle-of-the-country
vowels. As a child, I was maintaining oral family history;
I was tracking my family's lore.

"Maryland, stop talking like that," my mother said.
My mother trained my brothers and me to speak in
accent-less voice. She didn't want us to sound as nasal
and consonant-sloppy as the other Baltimoreans, so we
ended up sounding like we're from nowhere.

My red-haired friend Harp says that the nasal
Baltimore accent is southern but fast. "We're city

folks and don't have time for all the syllables," he says. Harp's a native Baltimorean, he's in three bands, and he sometimes refers to himself as the King of Peru. "The people on the Eastern Shore," he said, playing a chord on his base. "Now, they have plenty of time for all those syllables."

"And I don't have time to go to the Eastern Shore," I joked in a heavy Southern accent.

Maryland's Eastern Shore found time for me. In November 2002, the Dorchester Arts Center of Cambridge in the xenophobic, vaguely-Confederate, rural Eastern Shore county of Dorchester commissioned me to write a ghost walking tour for High Street in its historic West End. Formed in 1970, the Arts Center "is dedicated to providing and supporting art activities in Dorchester" and offers to the community "gallery shows, classes, art guilds, lectures and offices."[1] The Arts Center promised to pay for my travel and provide the research, and I feel sure I underbid everyone on the Eastern Shore.

My contact at the Center was its education and arts coordinator, Judy; she had a very slight Southern accent. She told me on the phone that she was "not from Cambridge but originally from the Eastern Shore."

I didn't hear from Judy again until late March 2003 when the grant money arrived and she called me.

"You have to meet the committee as soon as

possible," she said urgently.

"Okay," I said. "Tell me when."

I had never, as a Marylander, heard of any Cambridge ghosts, and, since I was a kid, I loved ghost stories.

Thirty years ago, my mother organized the annual Hampton Harvest School fair, and the highlight of the Fair was its Spook House, a 1960s ghost tour. It wasn't really a house; it was an underground crawl space the length of the building. Hampton Elementary School's built into the side of a hill, and its crawl space is a narrow, concrete path next to a hip-high wall and, beyond that, tightly packed, cool earth, sloping slowly up to the low ceiling. Its smell of moldy damp and decay creeped me out as a child; it stank of crypt and confining earth. I'm sure that stench contributed to the experience of our customers who walked along the darkened path, past the graves of monsters and the open caskets of the undead. My father painted cardboard freaks so realistic, that when we suddenly lit the fiends from below with big flashlights, the first graders screamed. My brothers and our friends dressed as mummies and popped out from behind the flimsy gravestones. We wrapped ourselves in rags and rolled bandages; we used red food coloring mixed into Vaseline for shiny blood. We laid out a smorgasbord of ghostly delicacies for the patrons to finger: peeled grapes for eyes, limp spaghetti for guts, gloves filled with wet sand for armless hands and Jell-O for gore.

Thirty years later, I was commissioned to write a ghost walking tour of one of the most prestigious streets in the state. I wanted to tell those stories. Bandaged like a mummy and crouched in the earthen dark, I wanted to be the story of the undead, to know the twisted history of the ghoul's previous life. What deviant path led to the cursed fate of the mummy?

Ghost tours have returned to popularity. Battle reenactments, haunted houses and ghost tours are all the American rage. We Americans occasionally grapple with our past and its inherent hyperbole, but, mostly, we're a teenage culture and just like being scared.

I took one of the Fells Point ghost walking tours to hear some Baltimore legend and to discover how much truth rippled through it. Fells Point is riddled with violent past; it was Baltimore's rough and tumble port neighborhood for years. The Point harbors enough phantoms to sustain two ghost tours; the one I took gathers every Friday at a toy store on Thames Street.

Fells Point has a carnival feel to it, even on non-festival weeknights. Its cobblestone streets are crowded with a brightly painted array of several hundred-year-old row houses and flocks of drunken locals and ambling tourists. Most of the narrow buildings house bars, so, despite its charming architecture, the stench of flat, grainy beer perfumes the sidewalks, even in the rain. It was raining that night but Baltimoreans didn't care;

it had been raining for weeks in March and we had become accustomed to the damp. I met my friend and fellow playwright, Kathleen, at Bertha's Restaurant for dinner, and over wine and mussels we updated ourselves on our various writing projects.

Kathleen was born in Baltimore, and we met in a playwriting class. She has sparkling eyes and is always impeccably dressed. Her curly, chestnut hair frames her pixie face. "Appropriate, you telling ghost stories," she said, grinning.

I wasn't too sure how to react. "Actually," I admitted, "I've never written ghost stories." Had I told her about the Hampton Spook House?

"How'd you get the Cambridge gig?" She asked.

"Young Audiences of Maryland referred me," I explained. Young Audiences of Maryland is a non-profit that supplies much-needed arts programs to the state's schools. "They market my . . ."

"The suffrage play, right. Cambridge. I don't know if I've ever been. Have you ever been?" Kathleen asked.

"Once or twice," I said, absently stabbing at my lettuce, trying to remember if my family had stopped at Cambridge on a houseboat tour of the Chesapeake years ago. In high school, I had hiked through Blackwater Wildlife Refuge in Dorchester County south of Cambridge. "Is it presumptuous of me to think I can write the ghost history of another town?" I wondered out loud. "I used to think that being a Marylander meant something."

"It does," Kathleen said.

"But what?"

"That we were all born in the same place."

"But the land in the state is so different. What does that mean?" I drank some water and stared out the window at the tourists wandering down Broadway. "That some of us know the state song?" I asked. The Maryland state song has the same tune as *Oh, Christmas Tree*. Most of the collective story that Americans share are Christmas carols and commercial jingles. We all can sing about the first Noel, Oscar Mayer and Coca Cola.

"Maryland, my Maryland," Kathleen said. We both hummed.

"This is my favorite verse," I said, singing. "Avenge the patriotic gore That flecked the streets of Baltimore, And be the battle queen of yore, Maryland, my Maryland!"[2]

Kathleen finished with me, laughing. The waiter gave us a funny look. "I don't remember learning that verse in school," Kathleen said, eating a bite of her omelet.

"It might be the first verse," I said, remembering. "I liked it as a kid because gore rhymed with Baltimore and yore."

"You would."

"Me and John Wilkes Booth. Apparently he liked

quoting it too."

"Not my favorite Marylander," Kathleen said darkly.

"I heard that he plotted Lincoln's assassination with the Catholic Church."

"Maryland! He was a Confederate spy and had nothing to do with the church."

"He was part of a conspiracy of Southern Maryland Confederate sympathizers that killed the best president in U.S. history. I like history."

"So, maybe you're perfect for that Cambridge job. Where will people walk on the tour?" Kathleen asked.

"They say it's a hike from the graveyard to the dock and back," I explained.

"Well, the graveyard has ghost promise," she replied.

"That's what I'm hoping," I said.

Later, as Kathleen and I walked through the light drizzle, I wondered at my literary arrogance. "The Eastern Shore's so different than Baltimore," I said to the back of her umbrella. "Almost like another country, another world. It's not just the Confederate Eastern Shore vs. the Yankee Western Shore but also the difference between country and city communities. How can I write it?"

"Really, Maryland, you're overreacting. They're people, aren't they?"

I shouldered against the crowd. Eastern Shore Tribal Council leader Chief Winter Fox once asked me about living in Baltimore, "Do you like living in the middle of all those people?" He lives in a thirty-person village deep within the southern swamps of Dorchester County.

"The store should be right on Thames Street," Kathleen said, forging ahead through the crowd and drawing me out of my thoughts.

"Do you find it vaguely creepy that a ghost tour starts in a toy store?" I muttered inside my windbreaker. "I mean, do toys bring ghosts? And vice versa?"

As we passed a knitting store close to the corner of Thames and Broadway, I saw a hard-looking man in black period dress with a long cape, seemingly out of place amongst the skeins of wool.

"That must be it," I called out to Kathleen, since this guy screamed ghost tour to me.

The man turned suddenly and threw me a very knowing glare, as if he was reading the lining of my xenophobic brain. His rugged face looked in its forties; his deep wrinkles burned into my skin. I froze under the fine rain, my bloated secrets spilt on the wet cobblestones, naked, wriggling and glistening. With a dizzying effort, I pulled away and reeled back. I looked up and saw the toy store sign further down the street. I swung back and the man was gone from the tiny shop. I blinked against the rain.

"Did you see that man, that man in the knitting store?" I asked Kathleen.

She had stopped under a dripping tree. "Man? Are

you flirting again? You write too much romance. Ah, there's the store," she said, pointing down the street.

Kathleen's much more pragmatic than I am. I didn't want to tell her that I thought I had seen an apparition. I was embarrassed and my shame clammed me up. Before this April, I would've needed a drink before I spilled my ghost stories to my friends. For the most part, they'd believe with me, but sometimes not.

When I told my friend Todd a ghost story, he tilted his salt and pepper curls and said, "Yeah, but you see things." Todd was born in Bel Air, Maryland and climbs a ladder faster than anyone I've ever met. I didn't quite know how to process his statement. It wasn't exactly positive reinforcement for ghost storytelling. Or maybe it was.

"Do I let myself see things that are already there?" I mumbled to myself as Kathleen shook her umbrella outside the toy store. "Or do I convince myself I see them?" My friend Laura says that she doesn't see ghosts because they don't believe in her. Laura was born in Baltimore and is smart and blonde and loves shoes. Her family's ancestral home was built on land deeded from Lord Baltimore.

I hoped to see the hard man later on the ghost walk, perhaps as a different tour guide. I hoped he was just another Fells Point freak in a cloak, yet the tingling fear of his chilling glance planted doubt. We eventually crossed paths with the other tour, but the dark, caped man wasn't guiding it. I was mystified why a grown man, such a hard man, would be skulking in period dress in a knitting store.

In the ghost tour toy store, Kathleen and I shopped for nieces and nephews until the damp tourists gathered to hear phantom folklore. One family had traveled from Australia. Our guides were in their mid-twenties, winsome and dressed in black. They told this profitable tour to over 600 clients during the summer of 2002 for $12 a pop with a little research and no overhead.

Our two young guides collected the group outside the store, and the stories began.

The tour moved as a loose pack, under scattered raindrops, hiking two blocks north to Friends Bar. As we passed an open bar door on Lancaster Street, the bouncer on his stool said something sotto and a male voice inside called out the stereotypical, wavering wail of a mournful ghost. The bouncer grinned and waved. The tour members giggled self-consciously, feeling sheepish about paying to hear spiritual folklore. The guides rolled their eyes and staunchly continued, shaking water and derision from their wet capes.

Many Fells Point ghost stories happen around last call in bars; you gotta like that in a neighborhood. Friends Bar regulars speculate that their ghost was a madam because of the midnight click of her high heels and her moans of bodiless passion. One tenant said she tried to pull him into his rumpled bed at 2 A.M., but I question his alcoholic state at that time.

The tour looped back to our dinner site. Bertha's Restaurant and Bar is a green-painted brick cornerstone of the wide square at the base of Broadway in Fells Point. The guide with the striped stockings and the fringed cloak of memory closed her eyes as she re-told a Bertha's ghost story. "The waiter opened the locked door and found himself face to face with a little girl in Victorian clothes sitting there, skipping rope," she said.

I frowned, trying to visualize that one. Raindrops rolled down my nose. I felt especially human and damp.

As our guides led us to the next stop at the Whistling Oyster, a bar across the wet square, Kathleen muttered to me, "How could the little girl ghost sit and skip rope at the same time?"

"I wondered that too," I said. "And how could the manager see ghosts on the video camera?"

Later, my friend Joan debunked the tour's phantom story of two spirit sailors in Bertha's bar, ostensibly viewed via surveillance camera by an invalid manager in the office. Joan tended Bertha's bar in the 1990s, and she and I met during a play. She's a Scorpio with beautiful hands. She can drive a tractor and write an elegant grant.

"There're no managers," she said, "It's just Tony and Laura, and there's no way that Tony would pay for a video surveillance camera of the bar. And even if the invalid was the manager, she couldn't get to the office with a broken leg; it's on the third floor."

The Fells Point tour guides used a little hyperbole.

Hyperbole builds tall tales. The legend of King Arthur evolved for five centuries before Sir Thomas Mallory wrote it down in 1469. In Arthur's early iterations, he's not a king. The growth of his story matches the evolution of the British Isles; his story grew bigger as its country developed.

We daily fall into hyperbolic traps. Each time a story is told, it morphs. My friend Adrienne said of a story that her husband tells of the night actor Bill Pullman checked her out, "Each time he tells it, Pullman does another double take. Eventually, he'll break his neck." Adrienne is a lighting designer from Long Island who taught herself to knit. She has creamy skin, and her hair and eyes are the color of warm mahogany. Each time her husband Lou elaborates his Pullman Tale that newly embellished version cements in his head until he begins to believe it himself. That's how legend evolves. We all perpetuate myth.

I am pretty regularly accused of exaggerating, but I'm a storyteller. I'm a flawed human source, and therefore muddied, muddy like a meandering river carving up an ancient marsh. Sunlight can't reach the bottom through all the suspended sediment of local and personal history. Maybe story can.

I had heard the story of the Friends Bar ghoul, and

I've seen first-hand the phantom drink at the end of the bar that the bartenders nightly leave out for their spectral madam. I've heard that the liquor vanishes each night, but it's an untended shot of whiskey in the wilds of Fells Point. Joan and I have heard stories about the upstairs Bertha's ghost and have felt an odd, cold, vibrating energy in the storage room where the waiter saw the skipping girl, a nearly palpable thickness in the air and the sense of being watched.

The tour's a mix of local legend and complete fabrication, and I'm not too sure where the demarcation line of fabulism wiggles.

Joan was born in Westminster, Maryland, but her mother's people hail from Kent County, north of Dorchester County on the Eastern Shore. She tried to explain the Shore separation to me. "They still haven't gotten over the Bridge being built," she said. The Chesapeake Bay Bridge was built in 1952 and shortened the trip from Baltimore to the Atlantic beaches by some four hours.

"The Bridge!" I replied. "They're not over the Civil War!"

"No, they're not," she agreed.

Americans are still defined by Civil War labels, and we like acting out our bloody history. I took my father to a Gettysburg reenactment where we watched General

Longstreet's fight against the Louisiana regiments. The reenactment was the biggest in history: 30,000 reenactors and spectators converged on southern Pennsylvania to try to resolve our nation's unresolved rift. One reenactor proudly told me that he was wearing his great-granddaddy's Confederate coat. The grandstands were packed shoulder to shoulder, and I felt terribly Roman as we watched men shoot at each other and dodge pyrotechnics in full uniform.

When Longstreet won, we spectators applauded. "Thank God, we won!" rose up a general cheer. We were genuinely happy. We are no longer split. We are still united. We are one nation divergent, under some God. We had to say it out loud, like a conjuring, like a prayer, to remind not only ourselves but the 51,000 phantoms hanging above us, right over our unsuspecting heads.

April 2nd

A CHILLED WIND

Because of its physiographic variety, Maryland is nicknamed Little America. The timbered Allegheny Mountains of the Appalachian Plateau in the state's western corner taper down into rich piedmont farmland and eventually meander into the coastal plain that engulfs the Chesapeake Bay. The roads between Baltimore and Cambridge belt cities, skirt suburbia and span a massive estuary. Half an hour and the city fades away as the road curves into the emerald tunnel of Route 97, banking south. A temperate zone has its price for green; lush vines and junk trees crowd the woods, cloaking light poles and road signs. Cross the Severn River twice, and hunting, fishing and boating shops crop up between strip malls. On the eastern side of the Chesapeake Bay Bridge, Route 50 sinks down to the level right under the horizon, where it bakes and floats all the way to the Atlantic. Rippling fields rise up around the tarmac, and the wind smells of brine and sun, even under nights of rain. The sky lowers over the car roof, and clouds touch down behind the farmhouses. Flocks of birds mysteriously emerge from the alfalfa and loop and curve overhead. The air's thick with water. The car hovers over the road. Time suspends to a crawl, stutters and corners of it reach backwards. People talk slower.

The ex-rector of Cambridge's Christ Church blames the town's many typhoid epidemics on its isolation. "Cambridge's so alone," Father Martin said. "They were all dying and had no where to go." Disgustingly, the typhoid outbreaks are better blamed on the county's extremely high water table. With a high water table, graveyards and privies easily contaminate water sources. Dorchester County is damp; its wetlands house a convergence of six rivers: the Blackwater, the Chicanacomico, the Choptank, the Honga, the Nanticoke and the Transquaking.

"So, you're going to Cambridge," said my friend Lou, grinning. At the time of that statement, we were sitting at the bar in the Club Charles in Baltimore.

"I've got a writing gig there," I said. "I know it sounds like the boonies, but there might be some good ghost stories." The real boondocks probably weren't Cambridge itself; the true boondocks lurked outside of town in the gangly stretches of corn, swamp and neck overwhelming it. "I think I wanna tell ghost stories."

"Nice," he said, tipping back on his bar stool. "Just don't leave Route 50."

"What do you mean?"

"Have you been there? You leave town and all bets are off. It's one freaked place," Lou said, smirking and drinking beer. "Cambridge's very strange," he said. "It's an odd mix. It's filled with people who think it's the hood, and with old, weird people and *Deliverance* outside of town." He's a mediator, another North Dakotan, married to Adrienne and the one who tells the Bill Pullman Story. He has short, sandy hair, and the friends in his group sometimes refer to him as The Emperor.

Late afternoon in early April, teetering on the brink of dusk, I drove southeast to my first meeting with the Dorchester Arts Center. Eighty miles from Baltimore and beyond the Bay Bridge and the Eastern Shore town of Easton, Route 50 twists at the edge of the thick forest of southern Talbot County, and a vista opens to allow the Choptank River, the northern boundary of Dorchester County, to cut its twisting path through the Eastern Shore to the Bay. As my car hurtled over the mile-and-a-half-long Frederick C. Malkus, Jr. Bridge, the setting sun lit a wide, orange ripple of water, rolling sideways across the darkening river, following me to the Cambridge side. The crest of the wave riveted me; surely it was an optical illusion or a boat wake. But there were no boats in sight. The river was clear, and there was no obvious source for the wave. I wondered about river currents.

Directly after the Choptank, I veered off Route

50's fast food sprawl and out of the 21st century onto Maryland Avenue. Victorian and Depression clapboard houses crowded out convenient stores and business parks. A deep, brackish drainage ditch lined the side of the road by the river; drainage ditches line most Dorchester County roads. Little earthen bridges over scummy moats allowed access to homes. After a mile of quaint Avenue, a narrow drawbridge spanned several hundred years and thin, idyllic Cambridge Creek.

"Once you go over the creek, you turn right into the town. You can't go wrong," said Judy.

The creek was thick with boat; sail masts hollowly clanked my arrival. Boat slips lined the retaining walls of restaurants and condominiums. A paddleboat named the Choptank Queen carried a smattering of tourists, pointing at the land. Seagulls screeched and swooped over the hood of my car. Several watermen unloaded dripping baskets off tipping oyster boats.

In the decades after the Civil War, during the unsteady Restoration, some Confederate Eastern Shore watermen broke stringent racial boundaries of the rural Eastern Shore. They worked side by side in the 1870s oyster industry with fishermen of color. Those color-less work habits drove Dorchester County gentry to complain that in Cambridge "mongrels were usurping the kennels of thoroughbreds."[3]

Like Maryland, Dorchester County is split socially

and economically, between the reigning, blue-blood elite of Cambridge's High Street and the fishing and farming, blue-collar marsh inhabitants outside of town.

"There's a huge disparity of classes in Cambridge, still," said my friend Debbie. She's a lawyer for the state of Maryland and a native Baltimorean. In college, she made up an adjective to describe the clanking, clicking noises that kitchen utensils make when they bump into each other; she called it *kirky*. Her slightly turned-up nose is coated with freckles.

"Weren't there riots in the 60s?" I asked. Those were the Cambridge ghost stories I remembered. "Something about the courthouse."

"It was the 60s, Mary. There were riots everywhere," Debbie said.

Cambridge has a population of 10,911 living residents in a wet, 6.78-mile stretch between a river and a creek. Its town dock was a battlefield during the Oyster Wars, and the Choptank was a battlefield during the American Revolution, The War of 1812 and the Civil War. In 1945, journalist Lee McCardell described the town. He said, "the old Eastern Shore is High Street, subdued, shady, possibly a little aloof, with fine old houses and spacious lawns guarding its dignified and stately march down to the river."[4]

That's High Street: aloof, stately, and guarded while four blocks away to the south hulks a clapboard ghetto of peeling, one-room shacks.

Wealthy historic High Street is a study in Queen Anne and Federal architecture, with some Greek revival, Georgian and Shingle thrown in to break up the columns. Primeval trees form an umbrella that shade governors, assemblymen and lawyers: the white men who once ruled Dorchester County and the state of Maryland. None of the original thirteen colonial families own the houses of High Street now, but they were once the first families of the county and the state. Their names are on the Declaration of Independence; their portraits hang in the Maryland State Capitol. I was told that there are thirteen families who still rule Dorchester County. Not all of the original names are the same, but shake the tree and the first thirteen thud to the ground.

The Christ Episcopal Church and the Dorchester County Courthouse cap the south end of High Street like porch columns; the Choptank River Long Wharf Park caps the north end. Cambridge Creek runs parallel to High Street from the Wharf to the back of the courthouse. Halfway down to the river, on the creek side of High Street, leans the Dorchester County Arts Center. The path of the ghost walk would span High Street from the Center down to the Wharf, up to the Church, across to the Courthouse and back to the Center.

Mammoth houses crouched back from High Street;

their wide, lush lawns provide a safe moat between the living room and the plebian path of the public trekking to the marina at High Street's river end. I had traveled over the bridge and through bubbling marsh to the true Land of Pleasant Living[5] during its gentrified settling into another bucolic night. High Street at April twilight could be the opening shot of an American dynasty film; the rich night poised on the brink of epic story.

These people are loaded, I thought as I unbuckled my seat belt. What's an arts center doing in this high rent district? The taxes alone could deplete a non-profit budget.

I'm thoroughly middle class. When I left my car, I immediately felt estranged, as if the Cambridge night oxygen was not suited for my lungs and as if the insects were whispering about me. The tree roots reached through the cracked sidewalks for my legs. A female figure turned into the draperies in the lamplight of #116. I felt watched.

The sensation of being watched is woven into the beginning of many Dorchester ghost tales.

I hesitated and forced myself forward.

The Arts Center at #118 High Street was built in 1790 and is half of what was once a hotel. In 1906, Captain James Leonard left the hotel business and sliced the building in two. Engineers rolled what is now #116 on logs to its present position fifteen feet away. Captain Leonard's granddaughter told me that her great aunt rode in the house as it rolled away, playing with her dolls.

116 from 118. It's a stretch, I thought, but cut the Divine Proportion in two.

1.618 or the number PHI or the Divine Proportion is the transcendent mathematical ratio that repeats in a staggering number of proportions across nature, ranging from humans to bees. PHI is so peppered throughout the plant and animal kingdoms that early thinkers deduced that the number's source was God. It surfaces in the spirals of seashells, tornadoes, waterspouts, crop circles and whirlpools.

What kind of building was left from a halved Divine Proportion? I thought as I stood on its rippled sidewalk. An unsteady, spiraling one, I decided.

The white-shingled face of #118 sported two letterboxes with quilting flyers beside its closed door; otherwise it could be a private residence. Bumpy bricks led to crooked steps. The front porch tilted towards the courthouse, and, yet, oddly, had a small puddle on it. Despite the rainy spring, I wasn't sure how a tilted porch could carry standing water. I opened the door and a bell rang, its sweet tinkling overwhelmed by the sound of whining sanders in the back of house. The air in the building sucked in, as if another door was simultaneously opened somewhere else. The gift shop door was to my right and a small art gallery to my left. The chewy air tasted thickly of wood. The gift shop was empty; its sand candles and watercolors yielded no clues to the committee's whereabouts. The whine of sanders that drowned out the doorbell drew me to the back of the house where a group of saw-dusted men carved duck decoys in a long hall. I returned to the front door and braved the shallow, narrow uneven stairs. My knees ached and popped as I climbed. The second and third floor hallways were lonely and twisted. The ceilings were low. Someone was watching. I turned, feeling vertigo, and the ceiling dipped lower. Disoriented and considering a retreat to the car, I staggered down the steps to the second floor hall where I stopped by a watermark on the wall above the wainscoting: a long, serpentine curve of brown, bubbled stain that I had not noticed on the search upstairs. I reached up to touch its peeling paint, and something cold and wet passed me, brushing by me.

Whoosh went a chilled wind.

Is it for me? I thought wildly.

Something sodden licked my cheek and splashed into my eyes. It burnt briefly. I wiped it away; it felt slippery, similar to glycerin. I lurched back to the stairs, running my damp hand down my jean skirt.

This will change everything.

Assume that death is a transition in dimensions. Assume that apparitions cross dimensions. What enables those moments? What allows certain spirit energy to cross over and leave only vestiges on my cheek? Once time parallels are linked is there a perpetual circle between them, a spinning portal of sorts? Like a spiral of the Divine Proportion?

"Do they need passports?" I asked the second floor landing.

Lit by the dim, front hall chandelier, Judy stepped out of the gift shop below, slight, freckled, and concerned. She looked like a substitute teacher in her pastel sweater and skirt. I stared at her, clinging to the balustrade, panting, wondering if she was real.

"Are you Maryland?" she asked brightly. She seemed real. She was way too perky to be a specter. Her cinnamon eyes matched her auburn hair.

I took a deep, calming breath. You can do what you want, I thought, but the outcome's gonna be the same.

"Who's going to be the same?" Judy asked as she led me through the gift shop to the back kitchen, a small, cluttered room lined with crooked shelves heavy with art supplies. Mismatched chairs ringed a table that was wedged between the shelves and the door.

"Where was this room before? When I looked before?" I muttered. "How could I have missed an entire room?" I couldn't hear the sanders of the decoy makers anymore.

"Here she is," Judy announced happily to two women in the kitchen. "I found Maryland!"

The two other women who comprised part of the ghost tour committee sat in the warm, cramped quarters. I knew intrinsically that these women were Episcopalian.

They were the image of the Episcopal ladies who had served me cool lemonade in my youth after service on The Church of the Good Shepherd lawn. I had met them at Easter vigils and potluck suppers. I had helped them serve Shepherd's pie and peas to the masses. But, Episcopalian or not, I was still a Baltimorean from the Yankee side of the state and, at the very least, an outsider.

I know a man who was born in Cambridge and, yet, despite his birthright, is considered an outsider by locals because his parents, who were teachers, transplanted there. I recently received a joke e-mail titled *North vs. South*; it listed advice for Northerners relocating South. A closing warning read: *"if you do settle in the South and bear children, don't think we will accept them as Southerners. After all, if the cat had kittens in the oven, we wouldn't call 'em biscuits."*

Among xenophobic Cambridge insiders, they refer to us outsiders as *foreigners*, pronounced *fo-nouners*.

According to Chief Winter Fox, the definition of a Dorchester native is one who can stand in the middle of the community and recite back six generations without consulting a crib sheet. A Dorchester native can visit all their ancestors' graves too, unless, of course, they've been swept away by flood, which is common. In 1608, the Nanticoke Indians recited to Captain John Smith their heritage of thirteen generations, back to the 1300s. Smith estimated that the Nanticoke tribe had one hundred members and the Choptank tribe had only thirty.

Thirteen and one hundred and thirty.

Thirteen, like the number of original Cambridge families, like the number of the original colonies, like the estimated 13,000 years that *Homo sapiens* have been roaming North America, like the 13.7 billion years that this universe has been around. Thirteen like the thirteen stars in Solomon's Seal, like the twelve apostles and Jesus, like the twelve Zodiac signs and the sun, like the thirteen arrows in the eagle's foot in the U.S. Great Seal, like the thirteen laurel leaves in the eagle's other foot, the eagle on the back of every U.S. dollar bill.

The committeewomen knew no High Street ghosts.

There's a ghost upstairs, I thought, but I bit my lip.

Lynn, the blonde Arts Center board member in chinos and a pale pink Oxford, was as flushed as her blouse. "Well, there's that Christ Church story about the slaves dying in the basement," she admitted slowly, leaning across the table, her eyes sparkling mischievously.

"But we can't talk about that," said Margot, the pencil-thin tourism director. Her bracelets clanked, and her nails were perfectly manicured. She wore chocolate-colored silk that matched her hair. She tapped her pen as she pinned me to my folding chair with her withering glare.

Of course not, I thought. This is Maryland's rural south. I could get strung up if I followed that lead. I sadly decided not to ask anyone about that story.

"It probably didn't happen anyway," Judy said from

the head of the table. "It's just a story."

"Stories are powerful," I muttered. No ghosts, much less slave ghosts, and I had expected a full list of research. The list was part of the agreement. I didn't want to dig. Digging took time. I needed the money quickly. Judy handed me a towering stack of history books to comb. They wanted to know when they could see a first outline. They wanted the tour up and running by mid-July.

"Is that possible?" Margot asked as she clenched her Mount Blanc pen.

"Depends on the amount of research I have to do," I answered carefully. "I think I was promised a list . . ."

"I'll call you with leads," Judy said quickly. "I'll send letters to all the High Street residents, asking them for ghost tales."

I thought of the bucolic, American upper class street outside and wondered how successful that request would prove.

"I'm sorry, but are you bleeding? On your cheek?" Lynn suddenly asked.

All three of them focused on my face. My left hand rose up and felt a raised, swollen, sore spot where the wall water had splashed me. I thought of asbestos abatement and Revolutionary War era lead paint. My fingers were damp with clear fluid, not blood. The fluid sparkled briefly on my fingertips.

"No, I'm not bleeding. I'm fine. Just a little

rosacia," I explained, hoping the embarrassment of that confession would distract them. I dug in my purse for a mirror, manically reaching for my makeup bag.

"You can go upstairs to the bathroom, if you'd like," Judy said. She looked worried. She stood and motioned towards the gift shop. Her nails were not polished. She painted water colors and threw pots with the art classes.

I'm not going back upstairs alone, I thought. It had gotten dark since I had swerved to miss a ghost in the upstairs hall.

I found my compact and checked my cheek. On my left cheek rose red bumps, like an old TB test, in a row along my zygomatic arch, the curved bone under my cheek muscle. It wasn't festering or open or bleeding. It looked like an allergic reaction, like poison ivy.

Get out, I thought.

"What POV do you want? Narrator?" I asked them instead, clasping the compact shut and stacking the books. They were baffled. "Point of view, POV," I said, trying not to shake. "Who's telling the story?" They hadn't considered that. They had spent all their previous meeting time deciding that the tour should be named *Tales of the Hauntingly Weird and Unusual.*

This the title of a tour of a street without ghosts, I thought.

"What does it matter who tells the story?" Margot asked, checking her gold watch.

"Well, it's folklore, so I assume that it's in a

storytelling kind of frame," I tried to explain. "Ghosts are more believable when you believe the person telling the story."

"Oh," Lynn said.

"Well, that makes sense," Judy said.

I left not long after, toting my tower of history books and back peddling for the exit. I felt vaguely claustrophobic in the building. Judy held the uneven front door for me.

"It rained," I said, stating the blatantly obvious. The street and the sidewalk were lightly dampened in the dark.

"It rains a lot here," Judy said wistfully.

The night air was cool and moist and blew the tomato blossoms on the plants that lined the porch. "Those tomato plants are so big," I said. Their yellow heads nodded sagely in the bug-whispered breeze. "Isn't it a little early for them to bloom?" The ones that lined my backyard fence in Baltimore were tentative green shoots, despite the rainy spring. These were in full bloom a month or two early.

"Oh, we need to thin those. They should be in the backyard, but things grow fast here. And we get lots of volunteer plants. All the water. Our basements flood monthly, but it's very lush. I can give you a cutting when you come back," Judy said. "I'm sorry about the lack of ghosts. I wish we had more stories for you."

Under the smell of flowers and rain, there lingered

the smell of history, of swamp, of decay, of rich story lurking in the night.

"Oh, the stories are here," I said.

I lugged the books to the car and waited for Judy to go inside. As soon as she did, I checked my cheek in the car mirror. It looked redder than usual. I dabbed a Wet Ones across it and took a Benadryl. I was proud to be retentive enough to carry Wet Ones and Benadryl. My eyes watered; I'm allergic to spring and its blossoms and pollen. At the Bay Bridge, my cheek stopped throbbing or maybe it stopped throbbing when I stopped thinking about it throbbing. I drove home through the night to read the history of Dorchester County.

I have written some romance novels and plays, and although a few have been historical in genre, I had never forayed into investigative folklore compilations. I was starting with the closed clam of the Chesapeake, the town that reluctantly relinquished John Barth and his cagey and brilliant *Floating Opera*, the town whose brick courthouse burnt down in the middle of the night in May 1852 when the fire originated in the Register of Wills.

The Orphans Court offered a reward for any information for what they suspected to be an incendiary event.[6] The night of the 1852 courthouse fire, Edward LeCompte, the Dorchester County deputy register and a descendant of one of the original thirteen families, took home his complete files on the Minutes of the Court. Local legend claimed that a disgruntled son started the fire when he learned that his recently deceased father had cut him out of his will. Historian Elias Jones records that Edward's father, Samuel LeCompte, died in January 1862, a full ten years after the fire.[7] No one was ever arrested for the suspected arson.

So the legend motive didn't apply. Did LeCompte have a premonition?

History does not always jibe. Its human record has few checks and balances, and Dorchester County lost most of its records that firey night.

I fell asleep with Jones' thick book on my chest.

I dreamt of a flotilla of boats sailing in slow formation down the Choptank, shore-to-shore, hull-to-hull. There was a festive quality to the boats. They were all painted bright blue and were draped in white Christmas lights. Dark lines connected the boats in a long, uneven row. Men in dark coats systematically hurled something overboard but I couldn't tell what. I couldn't see the nets behind the boats, but I could hear cannons in the distance. They were slowly dredging, dredging the river for a body.

I woke up suddenly. I must've jumped.

"Hey, babe," my boyfriend Karl said. He was awake, reading some heavy Holocaust tome, a different kind of ghost story. Karl was born in Washington, DC, and he wears one type of outfit: black tee shirt, khakis and scuffed black work boots.

I rolled over, away from his reading light. "If I disappear while I'm writing this Cambridge thing," I mumbled into my pillow, "the first thing you do is dredge the Choptank."

"Yeah, sure, baby," he replied absently, turning a page. He sounded as far away as 1852.

May 15th

PUSH BACK

"Cambridge's an interesting town," said my Auntie Tish. Tish was born in New York, is wonderfully tactful and has traveled the world. She used to live in St. Michaels, north of Cambridge, in Talbot County. "Not the most friendly of people," she finished, tilting her head and smiling sweetly. She would know; she's a minister's wife.

The Cambridge community was not pleased with the idea of a ghost tour. Town talk speculated that it would compete with the Cambridge West End historical tour, the Harriet Tubman Museum tour and the tri-annual senior bus tour. The organizers of other town tours felt challenged, and so they grew defensive.

"The West End tour coordinator won't share information," explained Judy over the phone.

"Well, then, I guess I have to read more," I said slowly. "All that research might lengthen the process."

"No one even believes that *Tales of the Hauntingly Weird and Unusual* will sell," Judy said sadly.

Then why not share? I thought.

On the other end of the phone, eighty miles and a world away in a city ten times the size of Cambridge, I didn't answer. I was the outsider with little say in the existing dynamic. I didn't want to land in the middle of a small-town, blue-blood tug of Episcopal war, but I had. Commissions settle one in the oddest of spots.

"We've had some pushback," Judy said, "but the Dorchester Arts Center's going to push on. We're going to tell the stories anyway."

What could compel them to rock the town boat? Then I remembered the profit margin in the Fells Point ghost tours. "Well, good for you," I said, impressed that Judy was so committed to documenting the folklore of a virtually record-less town.

In the absence of any High Street folklore, Judy gave me the phone numbers of two local sources: Thomas and Olivia. Olivia contributed a column called the Graveside Chat to *Revelations*, the Christ Church monthly publication that's chock full of information on the local dead. Judy's a member of the Christ Church congregation, yet despite her requests, Olivia's schedule couldn't accommodate an interview with us until way into June. Somehow, Judy thought that I, a *foreigner*, could persuade Olivia to chat sooner about the Christ Church graveyard.

Judy said, "Olivia's column is the first thing everybody turns to in *Revelations*, even when we get the copies in service."

I imagined a Cambridge congregation, dressed for spring, rustling in creaky Episcopal pews, shifting their collective weight and sneaking peeks at the dirt of the dead during some interminable sermon. I pictured a big headline that read: Foreigner Writes Local History! Church collapses in protest! Courthouse burns to the ground! Graveyard shifts two feet!

I considered the graveyard to be my primary source of ghost stories, so I called Olivia first.

"I'm sorry but you've called me during a Cambridge Woman's Club luncheon and my guests are leaving and so you're going to have to call me back in an hour," she gushed all in one sweet, Southern breath.

"Oh, I'm terribly sorry to bother you," I said, almost tasting sugar. "I'll call back." I hung up and shook my head.

When I called her back, she said, "I could make room in my calendar next week."

"Wow." I was surprised. I, the foreigner, had somehow persuaded her to a meeting before June. Maybe Olivia wanted to tell stories more than she cared about local politics and the dynamics of social pressure. Good for her, I thought. "That's great," I said, happily surprised. "I appreciate . . ."

"Well," she said, "I have another luncheon next week but I could do Tuesday. Where should we meet?"

"Um, I'd like to see the cemetery," I said.

"I'll see if Judy has a room available at the Center," Olivia said.

Didn't she want to go to the cemetery? Certainly, she didn't want to meet me at the Cambridge Woman's Club in their High Street house purchased from the Maynider family in 1922.

Judy's other source, Thomas Flowers, wrote a folklore compilation and a revisionist, elementary-school-level history of Dorchester County, a tome that Judy considered my primary source of Cambridge information.

"Everything you need to know about Dorchester County is in here," she said when she handed me the book.

The history book was nobly self-published: the typeset is Courier and set by hand. The margins vary. The maps are hand drawn. Half of page 55 is a hand-drawn sketch of a sea monster chasing a Spanish ship to America. Flowers was obviously an oral storyteller; I found grammatical errors in the book that only Huck Finn should commit. Or maybe Flowers was writing in the vernacular, like Mark Twain did.

Flowers was born in southern Dorchester County and has been a teacher, a principal and a county councilman. For years, he's been presenting folklore lectures around the county. The courthouse has a 1690 record of his namesake being fined; he's a native.[8]

"He might not be very nice," Judy warned telephonically. He was one of the protestors to the ghost walk. A High Street ghost tour might hinder his lecture circuit.

Thomas was polite when I called, but he said that he was too busy to meet.

"Got to get my garden going," he explained. His voice was slow, gruff and considered. "I'm in my eighties; gardening takes a lot out of me."

I commiserated. "The weather's been so cold and wet," I complained. "Still, I planted tomatoes last week." Silence; he didn't bite. "They look so small I wonder if they'll bear fruit," I blathered. Marylanders usually love discussing tomato plants; it's a statewide passion. I thought at least Western and Eastern Shore natives could share that. "The tomato plants at the Center look great. They're already in bloom."

But there was no tomato talk that day. "I can give you the address of a lady who's done my history bus tour before," he replied to my tomato pitch. I wrote down the address. "And there's another one, named Mabel, I think," he continued. I waited. "No, no, I don't have her exact address, but I can tell you that she lives on Locust Street in that yellow house with the iron gate. Do you know that house?"

"I think I can figure it out," I lied into the gap between us. I didn't know Locust Street, and he knew it.

I recounted the Thomas and Olivia conversations to Judy.

"I don't think Olivia's accent's that strong," said Judy. She told me again that she was "originally from the Eastern Shore." I realized that my accent imitation had crossed that Yankee line. I apologized.

"I am becoming aware," I said, "of the schism that splits this state."

"Oh, we're very different over here," she said.

I admit that adopting a slightly Southern accent sometimes makes life a little easier. It's amazing how a Southern accent will grease the wheel. Most people will open up to you if you have a tiny Southern twinge in your voice. Men hold doors for you and carry your packages. Customer service reps develop empathy.

Maybe that just happens here in Maryland at the fuzzy border between North and South.

The two women Flowers suggested I contact were senior citizens who conduct Cambridge historical bus tours three times a year. They told Judy that a daily High Street ghost tour would compete with their tri-annual business. They protested with passive resistance; they refused to "share" information, as Judy tactfully put it.

"There's no need to call them," she said to me.

"I couldn't call them anyway," I said. How could I call them with one address and a yellow house on Locust

Street? "They can't hide the history," I continued.

Until I re-visited Cambridge, I could only study. So, I returned to reading history books.

> George Calvert, Lord Baltimore, converted
> to Catholicism, and, because no Roman
> Catholics were permitted to hold public office
> in England in 1625, he resigned from his
> Parliamentary post. James I rewarded him for
> his public service by granting him large land
> estates to the northeast of the colony of Virginia,
> including the current states of Maryland and
> Delaware. Calvert's son, Edward Sackville,
> then granted land to all his friends and began
> the colony of Maryland.

Clearly, the English monarchs wanted Lord Baltimore to be a Catholic on the other side of the Atlantic ocean, so they tempted him with American land. They compensated his religious oppression with distant dirt and created my home state. A lot of Catholics followed Lord Baltimore to Maryland. Did the sweet tang of religious freedom make the swampy earth precious? Does the blot of that religious persecution still stain the land?

I lay in bed beside Karl, reading and wondering if I could stay awake during the research sections of the project.

I dreamt that I stood in the bumpy Christ Church graveyard, my back to the stone wall. A wind blew

my hair, and the door to the church flew open. Lord
Baltimore walked out, in full wig and frock coat. He
smiled at me and offered me a cigarette.

"I thought you were Catholic," I said.

I awoke as Karl lifted my hand off the book that had
fallen on my rhythmic breast.

"Was I asleep?" I asked ridiculously.

"Not for long. You can't read in bed. You're lousy at
it," he said, leaning over me to put the book on the floor.
"That thing on your face looks better." He landed back
on his side with an exhaled breath and returned to his
dictator volume. He was living on a grant to research
Stalin's tyrannical reign. We slept with books and
highlighters and left the lights on.

I trudged to the bathroom to splash my face with
water and check the damages.

The raised bumps where the water damage drop
splashed me were vaguely red, despite hydrogen peroxide
cleanings and Neosporin applications. The bumps were
no longer leaching clear fluid. In the busted capillary
beds of my cheeks, there was a slightly rosier curve, daily
less and less swollen, still a mark of the Dorchester Arts
Center, a branding of a foreigner in a strange land.

May 20th

READ FOLKLORE

I read Flowers' history book before his folklore book; that was a mistake because most of the ghost stories lurk in the latter.

I showed the history book to my *kirky* friend Debbie and told her the story of calling Olivia and Flowers as we drank coffee in her Rodgers Forge living room; Debbie's a caffeine addict. She serves coffee. The sun shone brightly through the blinds and etched a barred pattern on the Oriental rug. Birds chirped in the maple in the front yard; cars whooshed by. We seemed far enough away from marsh country to discuss its ghosts. I handed her the history book. "Check out the sea monster sketch," I said smugly.

"Oh my God, Mary." She giggled. "It's hand drawn."

"I know; bless his self-published heart. No ghosts yet, though. It's probably in his folklore book. I gotta find that. Guess I'll look in the old Waverly bookstores.

Aren't ghost stories folklore? And not history?" I sipped her strong coffee. "What's the diff?" I asked.

Debbie flipped past my marker in the history book and read the end. "I don't think there is one," she said slowly.

"I mean, is it all semantics?"

"Somebody made up both of them. Look. Here they are, Mary," Debbie said.

"How'd you find them?" I asked her from the couch. Reading the history book was a like learning German to me: slow going. I had trouble remaining conscious during the Byzantine renditions of Maryland land grants.

"Skip to the Civil War," she said. I've known Debbie twenty years. Her vocal tones were saying: skip to the Civil War, you idiot, as if to question, where else in this country's patchwork history is our greatest concentration of phantoms?

"Of course," I said, "The 600,000 ghosts of The Civil War. They were bound to surface."

"There's a headless ghost slave named Big Liz in the Green Briar Swamp," she said. She read aloud the directions to Big Liz's eerie corner of this world, south of Cambridge at DeCoursey Bridge. "She haunts a bridge off Route 50," Debbie said, shaking her head and laughing. "You pull over and honk your horn and she shows up, holding her head in her hands." Debbie knows Route 50 well; her parents own a Delaware beach house. "I don't remember seeing a sign on 50 for a dead slave girl on DeCoursey Bridge Road."

"Dorchester County doesn't sign much. They don't want foreigners like us driving the back roads," I speculated.

Big Liz's story inspired Debbie to want to go on a pilgrimage to Green Briar Swamp. "We have to see her, Mary," she said. "She's right outside town. We should go and explore."

> *During the Civil War, Big Liz's plantation owner, John Austin, forced her to bury some ill-gotten, Confederate booty in the densely-thicketed Green Briar Swamp. After she dug the treasure's watery hole, he decapitated her, and she still haunts the marsh and guards his abandoned gold. According to local legend, the treasure's still buried in the swamp. If you drive to the DeCoursey bridge, honk your horn and flash your headlights, then the wind will blow, and you'll soon hear her shuffling step. The car engine will stall as she limps into view, shoulders stooped, cradling her head in her arms, her eyes glowing. You're trapped, frozen, and can't move as she draws closer and closer.[9]*

"A swamp monster. That's more like it," I said. "She's definitely going into the tour." Still, I was a little peeved by this story. "Didn't Big Liz notice that Austin was wearing a tobacco knife big enough to decapitate?" I asked Debbie. "Did he always wear it? Did he bring it to hack through the brambles?"

"And if she was big, why didn't Big Liz fight back?" Debbie returned.

Instead, Big Liz exhibited a passive, abused woman streak that's more frightening than her haunting. Did she crave death? Was that option better than slavery? Was there resignation? Did she secretly love Austin? Hate him? Did he bury her headless body? He would've been a fool not to try, even in a marsh where the tides and the crabs daily vacuum up all detritus.

"Read the folklore book, Mary," said Debbie. "Find out. It's your mission."

"I'll report back," I promised.

The next day, I found the folklore book in a used bookstore in the Waverly section of Baltimore. I sank to the bald carpet and began to read beside the listing bookshelves. I was riveted. Customers stepped over me. Time flew by. The terrifying stories hailed mostly from just outside of Cambridge. The book's jammed with many rocking-good ghost stories, so many that before one tall tale a paragraph postscript categorizes the myriad varieties of Dorchester ghost story. According to Flowers, spiritual visitations are driven by: task, collection of body parts, message delivery, reluctance to depart, protection of treasure, or being trapped in a time warp.[10] That last one's my favorite.

On the crooked floor of the old bookshop, in my periphery I saw a small, black shape zip around a crowded shelf. I blinked. Did I really see that? Did it come out of the shelves of local folklore? I stood suddenly, shaking my long skirt, fearing a cockroach.

"Some people in this area are afraid to leave their homes at night"[11] reported folklorist Brice Stump about the residents of Bucktown, the closest village to the Green Briar Swamp. Green Briar Swamp is seven miles square of razor-sharp marsh grass and gritty, foul-smelling mud lined with thick huckleberry bushes. Some locals claim to have seen Big Liz's image emerge from the tidal waters under DeCoursey bridge. Fishermen have drowned there, in several feet of dark water. Some locals claim to hear gunshots over the marsh in the middle of the night and seen headless bulls, pigs and strange lights over the cattails. Hunting dogs won't go through the huckleberry bushes into Green Briar; they possess the sixth sense to remain outside its haunted boundaries.

Flowers' folklore book describes an extended version of the Big Liz story, embellished with motive. I phoned Debbie to elaborate. "I called to tell you the Big Liz back story," I said, feeling like a fledgling phantom storyteller. Big Liz discovered Austin's Confederate smuggling ring and reported it to Union agents, and Austin murdered her for the betrayal.

"Then why she'd follow him into the swamp, alone?" Debbie asked.

"It makes no sense. Did she have no choice? If she

didn't go, would she reveal her betrayal? There's more.
After the war, he died broke, and he never revealed
the location of the gold.[12] If I was dirt poor during
Reconstruction when everything had gone to hell," I
said, "and I knew the location of a fortune in buried
gold, I would've dug it up, ghost field hand or not."

"Why didn't he dig it back up right after he killed
her?" Debbie pointed out.

Something must have happened to stop him.

There's something murky about a swamp ghost that's
scarier than a drier one. Something about being soaked
in 140 years of swamp goo turns a spirit spookier. Big
Liz's headless specter is covered in slime, like the *Ghost
Story* ghost who hid under the lake for fifty years, waiting,
dripping, oozing.

Flowers tells a story in *Shore Folklore* of discovering
Big Liz's gold in Green Briar; he dug up a metal
box in the southern section that she haunts. Like
previous treasure hunters in what locals call "the devil's
woodyard"[13], Flowers abruptly was very aware of being
watched; he was surrounded by the eerie sense of not
being alone. He was equally suddenly "transported to
another part of the swamp."[14] He was somewhere else
sans shovel and hoe. He experienced a time and space
jump. I got goose bumps reading it. He ran until he
found his way out.

Flowers put himself into a story or he was testing the
story out; either way, it took guts.

"Anyone who goes into that swamp with a flashlight and a shovel is asking for trouble," said Debbie. "I'm not too sure I want to go now," she admitted.

"Ah, come on," I whined. "Who else will I get to go with me? Certainly not Karl."

"Certainly not. Maybe you shouldn't go at all."

"I have to."

Besides exploring the marsh interior, Flowers' recipe to summon the headless undead is to park on the bridge and blow the car horn once. Vernon Griffin, in *The Veil and More Folklore of the Eastern Shore*, recommends blinking the lights three times and hitting the horn two times to summon the ghoul. Folklore collector George Carey suggests lights three times and horn six times. Some versions name her Big Liz and some Big Lizz with an extra Z. This is how legend evolves and becomes localized. In sorting through volumes of Eastern Shore folklore, I discovered several versions of the same story. I suspect it's community-specific. I wondered how people conjured Big Liz before the car was invented. Make the horse whinny?

Dorchester County ghost stories have endured hundreds of years wracked with transformation. While the world outside the marsh whirled through change since the Civil War, Big Liz's phantom still haunts the Green Briar Swamp. A traditional narrative is stabilizing to an area, and we rely upon unstable oral tradition to keep it constant. The Big Liz ghost story is over 140 years old, and yet people still honk in the marsh and dream her headless shuffle.

Each generation revises story for its own needs. My childhood friend's fifteen-year-old daughter Ysabel lives in Easton, north of Cambridge in Talbot County. She's heard of Big Liz. "There's some ghost in Cambridge," she told me, "On a bridge or something." Ysabel was born in Santa Cruz and is wise beyond her years. She holds energy inside her slight frame and bounces a lot.

"Yeah," I replied, "the headless swamp slave girl. You blow the horn and she appears, holding her head, her eyes glowing. Then you can't start the car."

Ysabel's brown eyes widened. "Oh my God," she breathed. "I thought it was just a story."

"And every year some accident happens to the kids who try to conjure her. They drive off the road or get into some kind of car accident or they find themselves suddenly in another part of the swamp. Transported."

"I'm not going now," Ysabel said, twisting her hair. "No matter who asks me."

"I'd think not," her mother said.

Cambridge wasn't talking so I started asking my friends about their ghost experiences. My seemingly pragmatic designer friend John has one ghost story with a similar spatial displacement as Flowers' swamp tale. John's from Silver Spring, Maryland and most of his dramatic scenic designs have arches in the set somewhere. He's boyish and sweetly shy with occasional blue hair. Bored in a darkened rehearsal, I asked him to tell me a ghost story.

"I don't know what happened. I really don't," he said quietly. "I was walking with my girlfriend through her parents' neighborhood in Glen Burnie. She wanted to show me a spooky, deserted house at the end of a lane. The neighborhood kids had feared this house for generations."

A chain link fence surrounded the dilapidated house. John and his girlfriend slowly walked by the fence, towards the gate. It was summer.
"I don't know what happened next," he said, his forehead crinkling, still trying to solve the mystery of the memory. "I remember feeling pulled. Suddenly I was at the gate, with my hand on the latch, and something in the yard wanted me in there. The sound of my girlfriend's flip flops woke me, the sound of her flip flops slapping on the cement sidewalk as she ran for help."

"Wow," I said, considering the diabolical pull of the dead house. "Woke you."
"Yeah, it was like waking up and finding myself several yards away from where I remembered I had been."
"Freaky."

"It was years ago in college," he said,

smiling. "But I still don't know what happened." Like Flowers, he was transported through space and time through a worm hole of sorts.

I had heard tales of haunted, abandoned houses that trapped people, but John's was the first live account. All buildings hold energy; there's no doubt about that.

As she was leaving my neighborhood, my ex next door neighbor confided in me that "the old man across the street" once told her that at the turn of the 20th century the first three houses on the street, including mine, were illegal abortion clinics. A doctor had lived in my house and performed safe abortions in my basement. I stopped her story. "Whoa, I don't know if I wanna know this," I said, glancing at the back porch for a clue. "I may never do laundry down there again."

You can title search your house but you can never know what is soaked in the soil underneath it. Talk to your older neighbors. They're closer to death. Think of the stories they must know.

Read folklore.

May 23rd

FICTIONALIZED REALITY

Snug in my eighty-year-old, ex-abortion clinic of
a house, I slogged through the nautical history of the
Chesapeake Bay. I had read a line from the Oyster
War book three times without understanding when my
blonde, mermaid, mystic friend Korinne called from
New Hampshire. She's the one who saw the old man
ghost in my living room. Her interruption was a relief.
We discussed the tour and the origin of myth.

"What is a ghost story?" I asked her.

"It's a story about a ghost," she giggled.

"Thanks. The borders between history, myth,
legend, folklore and ghost story seem so sketchy," I said.

"It's a scale difference," said Korinne in her
forthright New England vowels.

"I've read Edith Hamilton and Joseph Campbell,"
I rationalized, "but I don't want to study too much
mythology before I write this ghost walk. They might
color my conclusions. I wanna try to sort through the

massive ideas on my own. The brain's the final frontier.
I should be able to figure out the answers."

"It'll take you longer," she finally said. "Years even.
No offense."

"No offense taken. I guess." I still struggled. "But
if I figure it out on my own, won't that be especially
archetypal? As a member of the group, individual
thought might actually be a collective metaphor or
concept of the group."

"Of course and that's the point," she said, a little
exasperated. "What if you started with the dictionary?
It's a good place to start. Find out what the dictionary
says and call me back."

"Wouldn't that be cheating?" I asked.

"More like cribbing," she replied.

According to Webster's Dictionary, a myth shoulders
this sweeping definition: *"a traditional story of unknown
authorship, ostensibly with a historical basis, but serving usually
to explain some phenomenon of nature, the origin of man or the
customs, institutions, religious rites of a people."*[15]

"Basically, some big stuff is explained through stories
over a long time," I said when I called Korinne back.

"And no one knows who started it or how it grew,"
she said.

Mythology might match history; it might not. Myth
begins with history, which might be true or might not,
and then it's layered with hyperbole anda believed by
alot of people.

"Myth and the Roman and Greek religions," Korinne said.

"The lessons of the Old and New Testaments are told in myth," I said.

The Oxford Dictionary defines legend as a *"traditional story sometimes popularly regarded as historical but unauthenticated."*[16] American legend tells heroic tales of George Washington and his cherry tree and of John Henry and the steam engine. Both myth and legend have an unverifiable basis, but missing from the legend definition are the big storms and the big-why. Legend is implied myth-light, yet legend and myth are often synonymous.

Both dictionary sources define folklore as *"the traditional beliefs, legends, sayings and customs of a people."*[17] Folklore can take the shape of anything from dresses to songs to stories to dances.

"If folklore explains the customs of a people, then myth can describe folklore," I puzzled through with Korinne, "But legend's part of the folklore list and legend and myth are synonymous."

"It's a logic equation gone way wrong," she said.

The English language embraces little logic.

Legend and myth require loose historical mooring, and repetition of the story and the embellishments by the storytellers smudge the line between documentation and manufacture. Folklore must be laced with some skinny threads of history; it wasn't created in a vacuum.

But folklore doesn't, by definition, require history; it can be completely fabricated.

"Total fabrication," I said.

"I hope that folklore's a mix," she replied. "What?" She asked when I laughed at her hope. "I hope there's some truth in it. There better be. John Henry did work for the railroad."

"George Washington was once a boy," I said.

"And cherry trees grow in Virginia," Korinne riffed.

"And Big Liz's plantation owner John Austin did own a plantation outside Cambridge during the Civil War."

Folklorist George Carey includes legend among types of folktale: fairy tale, joke, song, tall tale and legend.[18] There is no ghost story specifically in that list; tall tale might be its closest cousin. Like the missing, xenophobic, state nationalism word, the English language doesn't have an exclusive word for the tall tale that is ghost story. It's not surprising; ghost lore is mostly oral and not printed. The public doesn't read much of the written versions of folklore-developed ghost story; the public mainly reads completely fabricated ghost stories. How can we believe the fictionalized version and scoff the folklore version?

Simple. Fear.

Like myth and legend, ghost stories are told over generations, have an unknown authorship or the shared group authorship of a community, sometimes have a historical basis, and occasionally involve the phenomenon of weather.

"But ghost stories don't describe religious rites. Every day ghost stories . . ." I started.

"As opposed to Holy Ghost ghost stories!" Korinne laughed.

"Don't fit easily into religious doctrine," I continued, "unless the spirits are the poor souls trapped in Purgatory." Regardless of denomination, some people believe that ghouls are stuck between life and death, trapped in denial at the spot of their demise. "Big Liz shuffling through the swamp could be in Purgatory," I said.

"Sure, she could," Korinne replied, "But you don't want to think that dead Grandma banging around the attic or re-arranging the silver is in Purgatory! Grandma *can't be* in Purgatory!"

Purgatory in the Roman Catholic doctrine is a transitory state of punishment for worldly sins.

Maybe because of its lack of specific religious affiliation, ghost lore forces us to question our individual concepts of spirituality.

"Like Rebecca who doesn't believe in a god but believes in ghosts," I said to Korinne.

My friend Rebecca is a pale, dark-haired stage manager and the distillation of all that is ironic and sarcastic. She was born in Massachusetts and reminds me of a Charles Addams character. "I'm a very logical person," she once said to me, "and a card-carrying atheist, but I completely believe in ghosts. There's no question about ghosts."

> *She was on tour and calling a show from backstage of an old Philadelphia theatre. She was calling a complicated light sequence and she saw someone out of the corner of her eyes. She turned, thinking the person was a stagehand who had a question. She saw an older man with a sweet smile on his face and wearing an orange sweater, standing by the fly rail. As she opened her mouth to ask him what he wanted, he disintegrated, vanished into air. She kept quiet, afraid that the old union guys in the theatre would think she was crazy if she mentioned what she had seen. She realized that whenever she wore her orange sweater, the old man appeared and then disappeared. Right before the show left Philly, she finally asked the stagehands what was going on.*
>
> *"Oh, that's Babe," they replied. "Were you wearing orange that day?" Babe had died by the fly rail and wore a lot of orange.*
>
> *"I helped carry him out," said one stagehand.*

"You gotta believe in ghosts," Korinne said. "Well, you do."

"Maybe ghosts are memories come alive," I speculated. "Is folklore fictionalized reality?"

"Reality is fiction" Korinne affirmed. "And what's scarier than a once real ghost in an orange sweater that has the power to return?"

Regardless of the hair-splitting, myth, legend, folklore and ghost story are all subjective narrations of some reality, and their difference is that of process, how the account evolves, and that might create a different product.

Maybe folklore is the poor man's history.

I asked my friend Terri the difference between a historian and a folklorist. She put down the newspaper she was reading and said, "A tie." She's from Brooklyn, and she speaks her mind. Even when she's sleepy, she has fire in her eyes.

Written history is only slightly more reliable than oral myth. Read the newspaper; it's laced with mistruths. Every day corrections are listed. Such are facts. Every day we journal the world and argue about which version is closest to what happened yesterday. Japanese history books still don't mention the 1937 Rape of Nanking when over 300,000 Chinese were killed and 80,000 women were raped in six horrifying weeks. It's so horrifying that they can't acknowledge it. Regimes dictate history; the winners write it.

For as Winston Churchill said as he wrote his account of his war adventures, "History will be kind to me because I intend to write it."

Maybe there's no such thing as non-fiction. Everything is made up. Everything has spin.

Napolean said that history was the myth that man agreed upon.

There is no such thing as a completely objective reality. We are constantly all inside the personal film of our lives. We are producer, director, actor and audience for this constant film called reality. We all put spin on every single moment.

How can we honestly recount the adventures of our ancestors when we can't accurately and objectively report what happened yesterday? How can we recite six generations when the average American nuclear family barely has two parents?

Thomas Flowers, for all his charming vernacular, attempted objectivity. In his folklore book, Flowers vows to one source to tell her story "as close to what you told me as I could remember,"[19] which is valiant effort, but, like any folklorist, he can only narrate from his perspective with his voice. Each time a story is chronicled with a new voice it changes, hence, the different versions of the same tradition. Written folklore is the distillation of generations of gossip and the telephone game.

Perhaps we can't be objective about history because the human brain cannot distinguish between perception and memory. Scientists have mapped the area of the brain that sees an object and the area of the brain that remembers the image of that same object, and those areas are identical. Seeing an apple in front of us is the same as remembering that apple in front of us. We create reality in our heads.

I heard a wonderful legend of Columbus' arrival in the West Indies. He dropped anchor offshore and remained there for several days. The natives, who had never seen objects like the Portuguese ships before, could not recognize them, and, therefore, could not see them. A shaman, staring at the bay, noticed ripples around the ships' hulls and finally, after much concentration, saw the full image of the vessels. Not until he told the rest of the tribe could they also see.

Are there unknown objects, like ghosts and aliens, around us constantly, but we don't see them because we've never experienced them? Do we only allow ourselves to see them in places that we expect to confront them, like old houses, darkened swamps and dilapidated hotels?

The question is not whether or not the invisible world exists; the question is what stories do we create to describe it. I know there is another world, invisible to me now, like I know that my foot is below my chair yet I can't see it, like I can feel a basement under a floor. Yet, if we could see actual electrons spinning around us in constant, overlapping dust, we would shut down. If we could daily see ghosts in other dimensions, we wouldn't be able to function. Who could make it to work on time with dead swamp girls on Route 50? Maybe the gift of seeing the dead is a curse, and its truth is so scary that it's cloaked in story.

We're a species of storytellers, genetically wired to tell narrative, and religion and history are our biggest stories.

Webster's Dictionary defines history as "*an account of what has or might have happened, especially in the form of a narrative, play, story or tale.*"[20]

Maybe those who write history should be named *historytellers*. Why not? History tellers exist.

Maybe when the stories of religion meet the stories of history, we create ghost stories, and it's so big that one descriptive word can't contain it.

And it's so big that we talk about it for centuries and centuries.

Folklorist Carey spins a pirate legend that has survived over two centuries of turbulent change and has been recounted by at least ten generations.

> *Three Eastern Shore teenage boys heard rumor of pirate gold buried in the sand dunes along a river's edge. They spent all morning unearthing the treasure; the sand was not yielding. The sun was hot, and the boys sweated and complained. Finally, they dug out the line of the chest's lid. The lock fell away with the swing of a shovel. As they cracked the rusted lid, they were knocked back by a wind*

and the stench of decay. The ghost of the pirate
Blackbeard rose up over the dunes, spewing fire.
He was ten feet tall and wore a long, curved
sword. His untrimmed beard extended down
his chest and was plaited into ribboned tails.
He wore lit firecrackers in his hair. When he
laughed, a tooth flew out of his wide mouth
and his ropes of gold necklaces tinkled. The
boys fled and cowered in the reeds. After a
few shivering moments, they returned to find
no chest, no hole, just the blasé land and the
whispering pull of the river as it lapped gently
against the shore. No treasure, no tooth, and no
ropes of gold.[21]

As a historical figure, Blackbeard ratchets that ghost story up a level into legend or myth; pirates' tales are not all yarn. Blackbeard was real.

I called one of Judy's leads, a volunteer librarian in the Maryland Room of the Cambridge Library. Her name was Thomasine. "Judy said I could ask you some questions, if you don't mind," I started. I explained the project and asked about Blackbeard.

"Chesapeake Bay pirates are all myth!" Thomasine replied in a voice that pierced the air like a ship's whistle.

"Well, if you define myth as hyperbolic history," I argued.

Despite this classic Cambridge denial, it's not fabrication that pirates pillaged Chesapeake waterfront property for hundreds of years. In 1635, the first act of piracy was committed on the northern Chesapeake on Palmers Island, and from 1691-1715, freebooters or pirates so roamed the Bay that all waterfront plantations were built like small-armed forts. There were so many pirates that they impacted local architecture, and that's not myth. Several Dorchester legends tell tale of phantoms guarding treasure boarded up in secret panels, rooms and compartments in early homes. As late as 1779, the lower counties of Maryland were infested with pirates who daily stole boats, sheep and cattle. From 1780-1781, three picaroon attacks ravaged Vienna in eastern Dorchester County. But of the pirates who terrorized the Eastern Seaboard, Edward Drummond or Edward Teach or Blackbeard was the most legendary, and he often "wore loosely-twisted hemp cord matches, dipped in saltpeter and lime water, lit and slowly burning, hanging from under his hat"[22] that looked like firecrackers.

Most Maryland ship-related folklore focuses on sunken shipwrecks or shanghaied sailors but some is laced with themes of transition. As early as the 17th century, phantom ships have been sighted on the Chesapeake, and Dorchester County has its own ghost ship story.

Two men were fishing at twilight on the Nanticoke River near Roaring Point where a navigational light flashes to warn ships of a

shallow sand spit. As the sun set, a low fog
slowly climbed up the river through the mist.
The fishermen saw a trawler coming up the
channel; it had a red and green light on the bow,
a white light on the stern and another white
light in the center of the ship. The fishermen
heard no motor but the ship was obviously not
a sailing vessel. Still, it steadily sailed up the
channel. Horrified, the fishermen watched
as the ship headed straight for the sand bar.
They called out warnings, but the boat did not
respond. Instead of running aground, the ship
disappeared into the foggy line between the water
and the land.[23]

The spirit vessel crossed dimensions safely, regardless of physical impossibilities, and the warnings of the living fishermen fell on dead ears.

The Chesapeake has long been a violent battlefield. The British blockaded the waters of the Choptank and the Chesapeake during the Revolution and the War of 1812. During that later war, the British berthed in the Patuxent River and stole provisions, burnt small craft and often kidnapped able-bodied men as prisoners of war. Ships sank. People drowned. The Dorchester sand's paved with ancient coins, and Cambridge residents still dig for buried gold on nearby Golden Hill.

According to Judy, Thomasine has "dug up some doubloons" in local swamps with her husband and a metal detector. I asked Thomasine about her coin search.

"I never dug through the marsh for buried treasure! Who told you that?"

"Well, Judy had heard . . ."

"Only crazy kids go into the DeCoursey marsh to look for Big Liz, and they're usually drunk and get into car accidents!"

"Well, they probably need the liquor for bravery."

"And I never read anything about pirates or the hangings at the courthouse!" Thomasine continued in a high-pitched whine.

Hangings at the courthouse, I wondered. Who brought that up? And how wonderful. "When were there hangings?" I asked.

"I don't know dates."

I was getting nowhere fast. "Judy said that maybe I could give you a short list of topics to research . . ." I started.

"Not if it includes pirates!"

"Well, yes, pirates, witches, the Oyster Militia, dredgers poaching, Big Liz, any ghosts you got . . ."

"You're talking about myth!" She protested shrilly.

No, I thought, I'm talking about the hazy line between history and myth.

"Don't talk to me about ghosts on High Street! I have to work there!" She screeched and hung up.

Buried not too deeply under our seemingly solid sidewalks and parking lots is the blood-soaked, story-drenched soil of our not-too-distant history. We block out the secret life of the past. We forget that not too far behind every myth is a real life adventure. We forget that our own lives are ghost stories.

I was looking underneath the Cambridge sidewalks for something more, something missing, just out of my reach. I couldn't wait to interview Olivia and see the graveyard.

Getting no help from Thomasine, I read more history, digging for the gem of a ghost story. Regardless of its unreliability and its inherent fiction, our stories are our roots; they're our culture's childhood.

> In the 1870s, the oyster market exploded, and Cambridge and other Chesapeake towns like St. Michaels and Crisfield became oyster boomtowns, no different than the rough and brawling Western mining towns of the same period, rippled by arson, theft, murder and rape. Virginia watermen poached Maryland oyster beds, so the oystermen took to sea with Winchester rifles. Maryland created an oyster navy comprised of several steamers and fifty men to suppress the anarchy, but the Baltimore Sun reported that the bloated bodies of dead oystermen were clogging the Choptank. Full-scale battles raged on Maryland waters over oyster beds. When Virginia dredgers threatened to fire on Cambridge and burn it to the ground, the citizens organized the Dorchester County Oyster Militia in order to protect their oyster beds.[24]

Amazingly enough, despite all this violent nautical history and all the folklore about sunken vessels and hidden gold, I found no Cambridge oyster war ghosts lurking in High Street houses. Maybe like Nanking, Cambridge doesn't want to glamorize that part of its rough history with story.

I was reading at my desk and an elongated black shape whipped by along the baseboard, heading out into the hall. The phone rang. I blinked, and my heart pounded. Did I see that?

I answered the phone; it sounded real. "How was Thomasine?" Judy asked.

"She doesn't dig for coins in the swamps."

"Oh, I heard she and her husband did . . ."

"Nope. Or if they do, she denies it. Vehemently." I took a deep breath. "And she doesn't believe that pirates ever existed."

"Oh, dear. Well, they did. Blackbeard sailed the Chesapeake and murdered and stole and hid in the backwater of the Eastern Shore to clean his ships."

"I know. There were pirates in Vienna."

"The Oyster Wars on the Choptank happened right at the end of this street."

"I know. I read the book," I said.

"Maybe you could make something up. I've read your work, certainly you can make up a little something."

"I don't like the idea of fabricating," I said. "It seems like cheating somehow. Like all those fascist regimes that made up history."

"I don't think it's the same as all that. Well, if you can't find the stories, then you have to make them up," she argued.

I didn't want to see the logic in that, but considering that all of history is made up, I shouldn't have any qualms about embellishing some ghost folklore.

After a month and a half and no direct High Street ghost legend, I fabricated a short story for the tour on #115 High Street, the once home of Joseph H. Johnson, the man who is credited with writing the Maryland oyster laws. Inspired by the Dorchester Arts Center's gravity-defying porch puddles, I wrote that even in dry spells puddles appear in the downstairs hall and sometimes the puddles have a pristine oyster shell in the middle.

I wondered if one person could start folklore, like literature, or if folklore took a community. I felt guilty that I even considered writing a story of shanghaied oysterman rising out of the river grass by the wharf, bloated by watery death, dripping, hulking, thick with revenge. I wrote a puddle story for the tour, but I swear that all the stories in this compilation are as true as I found them.

This is all truth, as I know it. "It's the gospel truth," as they say in the county.

May 25th
BLACK SHAPES

The hard man in Fells Point was not the first ghost story in my life. I repressed many poltergeist experiences until they cropped up again during this project, surfacing like mysterious skin rashes, skirting around baseboards like the black shapes. American society does not encourage its members to tell ghost stories. Some spectral tales begin with a reckless group of teenagers telling ghost folklore and a phantom descends, as if the narration of the story summoned up the real thing. The fabricated stories of *The Woman in Black*, *The Ring*, *The Turn of the Screw*, and *The Mummy* all have curses associated with the telling of the tale.

I wondered if those curses turn the ghost stories dissociative. Joseph Campbell theorized that when a myth separates from the society that created it, the story becomes disconnected. Commercial television jingles and reality TV are decidedly dissociative; they replace the informing archetypes in our brains with a useless jumble of confusion. But ghost stories don't disconnect people from their society. They reflect our need as a society to know the Other Side of the Veil, to know what happens after death, to communicate with those who have crossed, to trick death one last time before he finally finds us, to cheat the Devil, and to transcend our own demise. Most ghost stories project a further reality than this one and disqualify the end of our consciousness. We need ghost stories. I find them more reassuring than the doctrine of most religions. Phantom legends have been part of our lore for centuries. The survival of these associative myths is a very positive cultural sign and a fictional Darwinism – ghost stories have common ancestors and the ones that best adapted to their surroundings endured. It's the natural selection of folklore.

By learning her stories, I was connecting to Maryland and to her checkered past.

I told my friend Joe the Big Liz story. I was evolving the tour's narrative voice by re-telling the stories and adjusting my writing style. Ghost stories, parables and myth are told in basic language since they've been repeated by generations. Folklore archetypes work like metaphors work: with something tangible they help us understand a disembodied concept.

"The wind will howl and soon you'll hear Big Liz's shuffling step," I drawled, telling Joe the swamp story.

Joe was born in Frederick, Maryland, and he owns

an antique store called Fat Elvis. He re-circulates dead people's stuff back into the community. "No more. Stop," he said. He paused and his brown eyes twinkled. "Okay, more. Tell me more. Is she carrying her head?"

That's precisely how I feel about ghost stories.

Stop. No. More.

We love ghost stories and repress them all at once.

"New ghost stories are scarier than older ones," Joe said. "Somehow the fresher ones are more powerful." His goatee flexed as he thought. "Like it can still happen now."

"Like the older ones are further away?" I asked. This thinking seemed so linear.

"Maybe," he said.

"But Big Liz is still off Route 50," I said dubiously. "Ysabel knew her story and she's fifteen."

The older ones hold power from all those years of the ghost being dead, from all those years of the story being told. Each re-telling by each generation makes the story more powerful; certainly each chronicling increases impact and increases the story's base. The reporting conjures it, perpetuates it, like prayer.

I took a chance and told him one of mine, just to keep the story alive a little bit longer. All this thinking on ghost folklore reminded me of a darkness I once saw, a darkness much deeper than the little black shapes.

"All right. I know a fresher, newer ghost story," I said

I was working tech theatre at Essex Community College in Baltimore and standing at the fly rail backstage during a performance of Camelot. A thick, black, lighting cable had slipped and hung in a long heavy loop, blocking the offstage exit of one of the castle units. The tech crew waited in the wings to roll the castle units offstage in a complicated scene change, and Beck, our beloved production manager, stood beside me.

"Go tell Eric that we're going to have dress that cable before we can clear the castle," he said. Then he slowly put his hand on my arm. "No, wait, you don't have to."

Rumor in the tech staff was that Beck was a warlock; maybe because he sported a long ponytail and could rip through lumber like butter. I don't know about his religious leanings, but he saw the darkness coming before I did, a moving darkness darker than offstage, gathering around the cable, not really lifting it, but pushing it up, up over the edge of the castle. All the crew techs saw it, and shook their heads, as if their eyes would work better after the shaking.

"There could be a logical explanation," said Joe.

"Maybe, but what?" I asked. "No one was up in the fly area. We all saw the darkness lift the cable up. What could that be?"

"Precisely," he said finally. "I said could."

"Maybe we want to believe the illogical because we

want to believe the illogical," I rambled. "But all of us together, having a group halluncination?"

"Isn't there a hospital right next to Essex?" Joe asked.

Essex Community College is adjacent to Franklin Square Hospital, and some of its theatre department folklore recounts the recently hospitalized dead visiting the stage. A good percentage of the theatrical family believes in ghosts. We're open to that sort of thing. In one theory, that belief is the reason that ghosts reveal themselves to us. We allow ourselves to see the unknown ships in the harbor.

My friend Tomi's best ghost story happened in a theatre. Tomi and I went to high school together. He has silver, wavy hair and a big belly laugh. He's one of our group's storytellers. Tomi's been seeing ghosts for years. "Have I told you my favorite ghost story?" He asked me, drinking his beer.

> *"We had finished the show and we were closing up Harbor Theatre. You know, Harbor used to be in that old Victorian building in Fells Point. North of the market."*

> *I nodded, smiling, remembering Tom Flower's Locust Street directions.*

> *"The stage manager went upstairs, right, to shut off the electric, to shut down the dimmers. So, you get that she shut everything*

off, electrically. That has to be clear. When she got back, we were all about to leave and we all heard that whoosh of a spotlight coming on — whoosh!

And a spotlight shone downstage center. A perfect, round circle of light right dead center. Now, there are some remains of juice in a circuit."

"Yeah, but not enough to turn on a spot," I said.

"And not a perfect downstage center spot."

"So, I guess you left," I said, grinning.

"Oh, yeah," he said, tipping back in his chair. *"We booked."* He smiled. *"Those people didn't want to stay for the real show."*

Not all theatrical types believe, though. I told my ghost cable story to two theatre carpenters recently. "Yeah, but you're nuts," said one.

"But that's the story how I remember it," I insisted.

Do you have to believe to see? See to believe? Do you have to believe in something all around you that's usually hidden? Is that faith? I believed the ghouls in the Hampton Harvest Spook House were real, even though we helped Dad paint them. Why do the young believe ghosts more readily than those of us whose senses have been dulled by decades of life? Are we "trailing

clouds of glory"[25] as Wordsworth wrote, losing our vision
of the mysterious magic of the universe as we age? Yet,
the older I become and the more I see of life, the more I
believe there's more to life than what I see.

I tried to remember the ghost in the Dorchester Arts
Council hall almost two months ago: the sensation of
being watched, the wet lick, the vague after burn, the
cold whoosh of air that passed through me, tingling my
skeleton and raising the hairs on my neck.

The redness on my cheek had mostly faded. The
bumps had lowered and turned white or clear. Weirdly,
the curve of the pattern has shifted slightly on my
cheekbone. It was inching up, slightly closer to my eye.
I'm worried that my body absorbed the wall water. If
the water caused an allergic reaction on the surface of
my skin, imagine the damage it could corrode inside.

Karl was afraid to kiss it. He kissed around the edges
of it. "Has it moved up towards your eye?" He asked.

"How could it?" I replied, turning over and away
from him.

That night, I dreamt that I was in a house, looking
out a window, at a graveyard. A man, tall, angular and
wearing a white suit, appeared in the graves. I knew he
was dead. I turned from the sight of him, and he was
suddenly beside me in the house. He was very handsome
and older with white, curly hair. We didn't speak. I saw

a woman in the graveyard, and then she appeared beside him. She looked like my dead friend Carol, all dressed up. I gaped. I waited for them to dissolve, to decay, but they did not.

The man said, "Everyone owes the debt of a death."

The woman nodded. "Everyone."

We dream in archetype.

I called Korinne and told her this dream. She laughed at its obvious message. "That was it?" She chortled. "That you're going to die?" The first time she meditated on her place in the universe her revelation was "space and time are completely irrelevant." She was furious; she wanted more. Then she said seriously about my dream. "Yes, it's true that everybody owes a death, but they also owe their own life. We owe a debt of life."

All these ghost stories are about the debt of the lives of its characters.

Each of our lives will be a ghost story.

May 26th

SHOW ME THE WAY

My friend Adrienne told me a story of a vision she had not long after her grandmother died. She awoke to her deceased grandmother, standing at the side of her bed and telling her to wake up and that everything was going to be all right.

"She seemed so real," Adrienne said.

Adrienne felt a great peace and went back to sleep.

"Was it a dream?" Adrienne wondered.

"You never told me that," her brother complained from the couch, looking up from a magazine.

We don't share our ghost stories. We should. Even if they are only dreams, they give us hope and shape.

We spend a third of our lives asleep and a quarter of that time in REM, the Rapid Eye Movement stage of dream. Is dreaming an unconscious resorting of our daily flotsam and jetsam? To Sigmund Freud, dreams were the direct expression of the unconscious. Simply put, we don't know what dreams are. They often seem like an alternate second reality, and many ghost stories have the tint of nightmare, especially when spirits and monsters appear without warning inside a seemingly normal reality. I wonder where I go when I sleep. I seem to journey, but I don't know how I find my way back. The mystery of it sometimes makes me almost scared to sleep again. It seems odd that an activity we need in order to live so closely resembles death.

I dreamt I stood at the edge of Cambridge's High Street Wharf. Cambridge calls it Long Wharf Park but it's really less of a wharf and more of a curved retaining wall between the city dock and the public marina. The breeze was blowing, and the clouds provided a low ceiling. A fine mist was lightly raining. I could hear the water nuzzle the wall and the slap of a boat moving quietly through the waves, but I couldn't see anything for the haze. I strained to see, but the river fog turned my eyes to water. I felt inside a cloud, surrounded completely by billowy wisps of vapor. I wiped away my tears, and out of the mist cut the bow of ship, slicing through the water towards me. I froze. It was a big ship; its curved bow was about two stories high. I was riveted by the vessel's progress, although I longed to turn away. The ship's frame was translucent and vaguely blue, and under it I could see the tops of tiny buildings and church spires in the Choptank. Very small people were standing on the tops of the closest edifice and waving. Firecrackers exploded over the river and their lights reflected in pulsing, watery windows. The massive ghost ship rolled over the perfect underwater city and steamed through me into the land. I was pinned beneath its transparent bow into the long grass. I couldn't move my arms; ship planks replaced my ribs. I sunk into the wet dirt and felt roots wrapped around me. I awoke in a clutch, sweating, my eyes leaking tears.

In my sleeping brain, I linked Cambridge to the ancient myth of Atlantis.

Greek philosopher Plato originated the slippery myth of the lost underwater city of Atlantis. Some say he overheard his mentor Socrates spin it at an all-night party. Plato wrote of a perfect society in an island city in the oceans off Spain that was destroyed by sudden flood. In every continent thrives the same saga of a colossal flood, the most popular being the Noah story of the Old Testament. There are approximately 600 flood legends, apparently told by the survivors who were scattered to all corners of the round earth. Maybe the flood myth endures because of its themes of retribution, cleansing and new beginnings. Scientists theorize that three massive floods occurred during the 8,000 years of the transition out of the Ice Age.

Humans have lost a lot of knowledge from the past. The Atlantis myth vanished with the Greek philosophers' demise and re-surfaced in the 15th century, some 1800 years later, when Christopher Columbus convinced Queen Isabella that Spain should track down and claim the legendary vast plunder of the lost kingdom. In his search for Atlantis, Columbus instead re-discovered the Americas.

That's the world-discovering power of myth. Thanks to Plato, we vacation in the Caribbean. Thanks to Plato, I dreamt of tiny cities under the Choptank waves, and life is a shadow on a cave wall.

Joseph Campbell said, "Myth is the society's dream."[26] Blackbeard and Big Liz are didactic nightmares that teach us greed, betrayal, and redemption.

Myths teach by example, and like parable, often that example is what *not* to do.

And, if we don't know the stories, we can't know their morals.

Campbell said that "one of our problems today is that we are not acquainted with the literature of the spirit" (myth) and we are ignorant of the "magnificent human heritage we have in our great tradition – Plato, Confucius, the Buddha, Goethe, and others who speak of the eternal values that have to do with the centering of our lives."[27]

Homer believed in ghosts. In *The Iliad*, dead Patroklos appears to Achilles in dream.

Shakespeare believed in ghosts. Dead Julius Caesar

appears to Brutus on the battlefield. Hamlet's murdered father shows up on the Elsinore battlements, warning about his evil brother-in-law.

Ghost stories sometimes quietly educate. A sad ghost story from Trappe, a small town on the northern, other side of the Choptank from Cambridge, has an alcoholic warning inside it.

> *The town's doctor was summoned late one stormy night to attend his best friend, who had been accidentally shot and was bleeding to death. The doctor set out with his horse and rig, taking along a jug of medicinal whiskey to fortify himself against the inclement weather. He drank too much and lost his way in the blinding sheets of rain. By the time he arrived at his friend's house, the friend was dead. In a rage of grief and shame, the doctor galloped wildly towards home, overturned his rig and died. Ever since then, he has roamed country back roads. On tempestuous nights, you can hear the creek of buggy wheels, the pounding of the horse hooves and the doctor's slurred voice, beseeching, "Show me the way! Show me the way!"[28]*

Missing from Flowers' ghost story list is the punished ghoul category, the Jacob Marleys of Purgatory, heavy lock boxes of guilt and grief dragging from their hollow waists. Add another phantom category: the restless, regretful spirit, cursed to atone for earthly missteps for all eternity. And what missteps the doctor made. Trappe's a diminutive community with few streets. Even now, you blink and you've passed it. How did he possibly become lost?

I reviewed my recent mistakes. I had cut my mother short on the phone. I had some road rage on Route 50. I felt a twinge of shame about the fabricated oyster puddle story.

I stopped working and paced to think. I compulsively touched my cheek. To my horror and relief, the watermark burn had faded into my skin. Was it ever there? Did I dream it? Could it have been poison ivy from my tangled backyard? I hadn't gardened in several days. No, it happened at the Center. I remember the whoosh. I searched my tattered skin for signs. The patterned curve of bumps had disappeared, leaving no pockmarks on the time-torn moonscape of my face.

Would this incident teach me to avoid the Dorchester Arts Center?

I doubt it, I thought.

Show me the way.

The ghost bumps vanished the same day that two sheets of Christ Church graveyard research disappeared. Judy had given me a few pages of research on the sad life of Willimina Smith Goldsborough, whose poignant story of marrying the wrong man qualified her as a candidate for the tour's narrator. Willimina's love story isn't a phantom one, but her lovelorn voice has tinges of remorse appropriate for ghost stories. Her research pages were in my Cambridge file and they've gone missing from my office, as if my narrative voice had been kidnapped. I must've misplaced the pages. I tore through my office and re-sorted many stacks of paper, but I only raised a cloud of dust and sneezed.

They're gone.

People have gone missing in the wooded bogs and vast wet savannahs of Dorchester County, never to be seen again. It's not surprising and, actually, almost expected. Southern Dorchester County roads curve through thick swamps with no guardrails. One wrong turn and you're sinking into the gluey mud. Marsh residents tell several vanishing stories about natives who insisted on living alone, surrounded by quagmire.

> *Two unrelated people disappeared exactly the same way from the same community seven-years apart. Seven years and practically the same story was told again but with a different protagonist. Both victims were the last of their family lines and lived alone in a house at the end of an empty lane down around Bestpitch. Both investigations uncovered their tracks into the marsh beside the dead body of a family pet (a dog for the man and a cat for the woman) and a tidy pile of kindling. The footprints just stopped, leaving behind Egyptian-like transition*

tokens of companion and fuel.[29]

How freaky is that? An old lady vanished and seven years later an old man vanished exactly the same way: tracks that ended in the marsh next to a dead pet. There's more.

> *An African American man vanished*
> *off a muddy road outside of Aisquith. He*
> *had a limping, stuttering step and his muddy*
> *tracks stopped in the middle of the road, as*
> *if something had lifted him straight up in the*
> *middle of the woods. Locals said that saucers*
> *got him.*[30]

A third of Americans (or one in fifteen) believe in extraterrestrials, and five million of us have reported UFO incidents, many in upstate New York. In a 2002 Roper Poll, seventy-two percent of adult Americans stated that they believe that the U.S. Government is keeping secrets about UFOs. Realistically, earth is thousands of light years from the closest possible, viable, life-sustaining planets in our 200-billion-star galaxy and at least thirty light years to our nearest neighbors in the Milky Way. It's almost as lonely as living in a swamp.

I wrote a science fiction play with my friend Tom. Tom's a native Baltimorean who's an oxymoronic combination of grace and awkwardness, of conservative and wacky. He's tall and long-boned and laughs easily. I wrote the comedy in exchange for a week's vacation in his family's rustic cabin on a private Adirondack lake in upstate New York. In the play's plot, aliens from the Dog Star Sirius plan a takeover of Earth; that was Tom's idea. We staged the play at the lake's tiny community cabin.

> *After entertaining the lake vacationers with the alien play, a dozen actors drank on the end of the pier back at camp. Suddenly, a triangle of red and green lights zipped over the treetops. The triangle ship danced back and forth, almost balletically for us. Sometimes the lights would go straight back, out of our perspective, and re-appear just as suddenly. When the aircraft disappeared, it seemed to travel away from our depth, as it moving into another dimension. At the end of the display, the lights flew over the trees and, in a sudden burst of speed, completely vanished. I stood at the end of the pier and called a drunken thanks to the heavens. The actors dispersed, ostensibly to fetch more beer. When questioned about the extraterrestrial event at breakfast, the group denied any alien light show, but throughout the rest of the day, individuals admitted quietly to me that they had seen the unexplained dancing flight as well.*

> *Tom said, "Not everything means something, Maryland."*

> *I think he might be wrong. Something that big should mean something.*

Many of us don't like discussing the unknown when

it affects us directly. Take, for example, the city of Cambridge.

Dorchester County locals regularly report odd lights over the marshes: round, brightly colored lights that hover in groups or shoot straight up into the night. Mr. Travis saw green lights over the Aisquith marsh. Most of these lights are the result of natural gases rising out of the swamp, but locals say they are warnings of death.

Folklorist Harold Roth recounts a light-in-the-sky story from the Hill family of Cambridge.

> *The Hill family lived outside Cambridge and was sitting at dinner one day when the mother saw a bright light flash through the sky. Everyone else at the table followed her stare and saw a trailing glow, brighter than the sun, dropping through the blue, cloudless firmament. The spectacle fell behind a stand of trees and landed with crash so severe that it shook the family's china.*
>
> *The mother said, "It's a token of something dreadful to happen soon."[30] Within a half hour, the news came that the neighbor on the other side of the trees, in the direction where the light fell, died at his supper table, about a half hour before the family saw the light.*

This folklore grants me great consolation: someone or something comes for us at our deaths and we turn into light. That radiance came for the neighbor, and he didn't cross alone. It was his fate. Humans assign intent to ghost stories; we need a reason why we're alive. We long to have a destiny, and we cling to a John Irving-esque fatalism, that our every action changes not only our course but the course of history and that, no matter how we struggle, we are not self-determining and can only follow our destined path. It's so much easier to live that way.

May 27th

GHOSTS OR NO GHOSTS

I seem destined to leak. Fluids were sneaking out of me. I lay on my side, watching the History Channel, and a lone clear tear oozed out of my right eye and slid to the pillow.

"Are you crying?" Karl asked. We were curled up on the couch, watching a program about the creation of the national cemetery in Gettysburg.

I sat up abruptly and kicked his leg by mistake. "Sorry," I said. My eyes had been tearing; water escaped from my body through them. There was a constant film over them. "Maybe there's a high pollen count this spring," I said. I had been sneezing a lot. That must be the logical explanation.

Spring exploded into a raucous pageant of deep color: magenta, pink and red. Flowering trees scattered blossoms like snow, and pink and white dogwood and flowering crabapple dotted thick banks of new green.

Cottony clouds checkered the pale blue sky above Baltimore. On my second journey to Cambridge, somewhere around the Bay Bridge, the warm wind blew them all apart, and the hint of sea air curled like smoke into my lungs. I drove in the right lane over the Frederick Malkus Bridge; I wanted to be closer to the water. The river glistened beneath my wheels, but no wave carved its surface. Gulls greeted me with nasal squawks.

Olivia wanted to meet me at the Dorchester Arts Center. I had no desire to return to the building that branded me, but I had little choice. I consoled myself with the concept that haunting is not constant. Even if a spirit occupies a site, it's not always there, no more than I am always at my home.

I was late for our mid morning appointment, and I felt bad about keeping Olivia waiting. I couldn't tell if she was aggravated; it's sometimes hard to read Southern women. We shook hands; hers was tan and limp. Under layers of lipstick, her smile seemed tired. She wore flower-printed pedal pushers and gold jewelry. Judy sent us upstairs to the conference room across the hall from the water stain. I tentatively crept up to the second floor landing, quietly quivering. The stain was still on the wall but oddly faded, as if it leaked years ago. Olivia sat primly at the six-foot folding table in the conference room, her roasted hands folded. She was born in Worcester County on the southern Eastern Shore. About fifteen years ago, the Christ Church congregation asked her to write about its graveyard, and since she was interested in genealogy, particularly in that of her husband's family, she agreed. Olivia's very sweet and seemingly delicate. Once a month, she encapsulates the life story of a graveyard occupant into two or three paragraphs. She handed me several documents, chronicling the best residents of the boneyard. I skimmed, hungry for ghoul.

"We have no ghosts, but we have some oddities," she flatly stated. Her shining bracelets clanked like the ones worn by the head of tourism.

"No pirates, no shanghaied sailors . . ."

"Oh, I don't know anything about that."

No ghosts inhabiting a several-hundred year old cemetery. Right, I thought.

The graveyard houses the oldest grave in Dorchester County (Magdalen Stevens who passed away in 1678) and is a crooked grid of Revolutionary War veterans' monuments and the remains of four governors and dozens of Confederate heroes. The church burnt to the ground in 1882, and many parochial records were lost in that fire. Scads of unmarked graves, whose wooden crosses were swept away by flood, riddle the yard. Gravediggers often hit unmarked brick vaults when digging a new hole; sometimes they hit a cradle vault of an infant typhoid victim. Sometimes they hit water and lots of it. The water table is high and close and filtered through hundreds of years of dead bodies. Some burial vaults are trapped under the paved stone of High Street; the graveyard boundaries have moved in its 280 year history.

"Sometimes during vestry meetings, the church makes odd noises," Olivia admitted with the memory of a smile. "They say it's just an old building but I joke that maybe it's the Reverend Daniel Maynider, who might be buried under the nave."

Fifteen years of research and married to a card-carrying native member of one of Cambridge's thirteen first families and Olivia still can't uncover all the church's mysteries. Why do I think I can?

"Have you ever heard of Hannah Maynider?" Asked Olivia.

I had read about her.

Reverend Daniel Maynider, Jr. was Christ Church's rector in the 1760s, and his mother Hannah lays claim to the best ghost story in the region. Hannah's true maiden name and her date of birth are debatable and one folklore version assigns her a different first name, but there were two Reverends Daniel Maynider, father and son, and there lies the confusion.

"Is she in this graveyard?" I asked.

"Oh, no, she lived across the river in Talbot County, near Trappe," said Olivia. "She was buried at St. Peter's; the road's right off Route 50."

Hannah fell into a coma and seemed to

*die. According to Olivia, Hannah had told her
husband that she wanted to be buried wearing
a treasured family heirloom, a huge ruby ring.
According to Flowers, the weather was warm
when she fell into her coma so her fingers were
too swollen to remove the ring. One version
says that as she lay in her open casket viewing,
thieves noted her fine jewelry. The Trappe
version, that refused the notion that the thieves
were local, says that two out-of-town sailors
heard of the ring in a bar. Regardless of how
they accumulated the jewelry knowledge, the
grave robbers dug her up the night she was
buried to cut the ring off her finger, but they
couldn't open her hand. One version says that
the fresh air revived her when the thieves pried
open her casket; another version speculates that
the pain of the cut yanked her from her coma.
In any case, Hannah awoke during her own
grave robbery, as the thieves' knife was pressed
into her skin. She screamed enough to shake the
church windows, and the thieves understandably
fled. Hannah dragged herself out of her early
grave, stumbling on her shroud. She staggered
home to her shocked husband.*

*It's Mark Twain's The Golden Arm meets
Ira Levin's Deathtrap in the late eighteenth
century, and I found three slightly different
versions of it in folklore books.*

*One version claims that when Hannah
collapsed at her own doorstep, her cut arm left
a bloodstain on the doorjamb that could never*

be cleaned. Even painted over, the stain crept
through, like truth peeping out of the whitewash
of history.[31]

The story is not that improbable. Live people are still misdiagnosed as the dead; sometimes even now we can't identify the mystery that is death. I read that a man in Raleigh, NC was declared dead at the scene of a motorcycle accident, and two hours later woke up in the morgue. Enough poor devils were buried alive in the 19th century to develop a new graveyard product by 1882: bells rigged into caskets and tombs. My great-grandmother Nannie insisted that she be cremated because she knew a woman in turn-of-the-century, rural Illinois who awoke from a coma as the dirt hit her casket.

"Let me out!" The woman cried, banging on the lid.

My younger brother remembered Nannie's buried alive story differently. He remembered the story ending with the woman's suffocating death, not her redemption. Our two versions are a clear example of how folklore evolves, even within the tight confines of the immediate family. Or maybe I'm claustrophobic and remember the better ending.

Humanity needs these stories. It's simple; we want to conquer death. Death and resurrection archetypes are stamped all over Hannah's story. Most ghost stories are riddled with archetype, if you believe Carl Jung's theories that all mankind carries common metaphors imprinted in its neurological hard wiring. Early man fabricated archetypes from his observations of reality. Even before the development of language, man fashioned symbols, and those symbols or archetypes are an extension still of our reality, part of the reality that we all produce. We share them. We share them in dream. We share them in the same way that every person on the planet knows how to turn a shirt inside out. Archetypes are woven through ghost stories and the unreliability of history. A woman rising from the dead and dodging her tragic bond with death is clearly one.

The Jungians argue that the gods and goddesses of myth, legend and folklore are archetypes, real possibilities in the brain, and if we encourage and explore those prototypes, we can develop into happier humans. It's easier to categorize information if you know the archetypes; you can recognize the symbols. You begin to comprehend why the symbols ring like clear graveyard bells in your heart and head.

"Why do people like ghost stories?" I asked my friend Mark. He's from North Carolina; maybe he knew the guy in the motorcycle accident who awoke in the morgue.

"They like being scared." He answered definitely, immediately. He's very smart; he's usually sarcastic but that answer wasn't.

"They want the safe fear of the roller coaster,

the thrill without the danger," I said. "They want confirmation of the beyond, that consciousness will not end. They want a message with no complete answer, a doctrine without the responsibility of a religion, instant karma."

> *The 1920s Dorchester County Shiloh*
> *Camp meetings often had as their star attraction*
> *Sarah Jones, a woman who had returned from*
> *the dead. Many meeting-goers believed that*
> *Sarah's touch could protect them from their*
> *earthly bargain of temporality. One touch*
> *from Sarah and the redeemed would live forever.*
> *Considering rural medicine at the time, Sarah's*
> *story's not atypical. Like Hannah, she fell*
> *into a lingering coma after a severe summer*
> *fever. When her relatives could no longer see her*
> *breath on a mirror, she was washed, cleaned,*
> *dressed and laid out for a three-day viewing*
> *with tubs of ice below her bier to slow decay.*
> *At her funeral, when the coffin-maker spread a*
> *cloth over her face, he noticed the fabric moved*
> *with her breath. A calm, quick thinker, he*
> *swung back and slapped her. The congregation*
> *gasped in shock but was more amazed when*
> *Sarah awoke and spoke her mother's name.*
> *Sarah never married, and she traveled the camp*
> *meeting circuit for the rest of her life, telling her*
> *story of her visit to the Other Side.[32]*

I wish I could have heard her testify. What story did she tell? What did she see while she slept? Did she ever go down the white tunnel to the light? Was she told to return? Flowers said she thanked God during her testimony in a "catatonic-like subdued retelling."[33]

Near death experiences (NDEs) have strikingly similar stories: an acceptance of death, an out of body sensation, of hovering above the dying body, an expansion of consciousness, an overwhelming feeling of bliss, a rapid movement through a tunnel or vortex to a beautiful white light, and a calming being of light who offers the option to return to the living. The experience is usually lacking any fear or regret.

NDEs could have purely physiological explanations: low blood oxygen, blocked brain receptor sites, or endorphins.

But why tell the same story over and over again?   It must be archetype.

"We tell ourselves stories in order to live,"[34] author Joan Didion wrote in the 1970s. Or rather we tell stories to assuage our towering fear of death. Stories can teach, warn and entertain. We tell stories every day and most of them justify ourselves or chronicle the reality film that we live. Some chronicling stories do more than relay information; sometimes we tell stories to establish reality. Ghost folklore does that: interpret and establish reality.

I don't think I could get through a day without the justification of a story or a joke.

Jokes are included in George Carey's folklore list, and my friend Lisa told me this one. Lisa was born in

Bel Air, Maryland and is a beautiful dancer who grows her own basil for pesto. She's a great hostess and her chin tips up when she talks. "How does one WASP propose to another one?" She asked me, her voice husky with irony. "Would you like to be buried with my people?" I laughed and laughed. "It's not that funny, Maryland," she said.

"Oh, yes it is. I can laugh; I'm a WASP."

"Did you ever hear of Sarah Jones?" I asked Olivia in the Arts Center conference room.

"I don't believe that she's in the graveyard with us," she replied in true WASP fashion. Olivia has catalogued the whole graveyard, notating all the epitaphs, except the unmarked graves and "a few colored folk." Olivia likes the graveyard stories of sad women; she associates with them. Ann Weller's grave is completely overgrown by a tree.

"Willimina Goldsborough died of a broken heart. She followed her father's wishes and didn't marry her true love," Olivia whispered. She folded over in her folding chair to confide in me, as if she was gossiping about the living in the next room or down the street.

"I heard about her," I whispered back, leaning in. "Very sad. I thought she'd died in childbirth."

"Her heart just gave out," Olivia said wistfully.

"Judy gave me her information," I confessed, "but, I think I, um, misplaced it."

"I'll get you my copy," she said. "I have a longer version anyway. She married a Goldsborough, you know."

Names and stories define people. We carry our favorite stories around with us, and they sink into our hearts and hands and become part of us.

"They lived on Horns Point where the college is now."

"Horns Point. That sounds familiar," I said, wildly taking notes.

When Olivia told me Hannah's Lazarus story, she described the Reverend Maynider as "a Huguenot, you know."

The Huguenots fled France's religious persecution for the spiritual penal colony that was Maryland. Following in Lord Baltimore's Catholic footsteps, in the late 17th and early 18th centuries, the English courts expatriated Roman Catholics to Maryland as punishment for their religious beliefs. In addition to those sacred exiles, according to the Flowers' history book, flocks of escaped indentured servants burrowed into the swamps and lived on to populate Dorchester County, the closed Australian clam of the Chesapeake. According to a fellow from Aisquith Island, descendants of Revolutionary War deserters populate the area around Blackwater Wildlife Refuge where they avoided Maryland and Virginia's enlistment laws. William Vans Murray of High Street was the chief of Scotland's

Murray clan until he fled to Maryland, following the ill fated, 1715 Jacobite Revolution.

A swamp's a good place to hide.

My friend Joan told me that Kent County residents don't care much for the southern Eastern Shore counties. "We're blue blood. They're swamp trash," she said, confidently, earnestly. Kent County is two counties north of Dorchester and its lower half is geographically marsh.

Certainly the Cambridge blue bloods have no desire to remember or recount their cloaked, Catholic, war-deserter history. They'd rather concentrate on the regal heritage of the Lord Baltimore land grants, property in huge chunks that Baltimore deeded to his friends in order to make the land productive. He awarded huge territory parcels and, in return, his friends paid Baltimore rent in tobacco. Some Cambridge natives proudly trace their lineage to those first land grants. The largest grants, the ones over 2,000 acres, were legally constituted as manors, and there is still debate amongst historians as to the degree of the transported English manorial system in Maryland.

The people of Cambridge really thought I'd believe that the state that transplanted and maintained the British manorial system in America has no ghosts, that the county that harbored a variety of exiles has no phantoms, that the town that waged war with Virginia oystermen has no spectral vestiges and that the street that houses the oldest graveyard in the county has no haunts.

Please, I thought.

 "Call me if you have any questions," Olivia said sweetly, as she gathered up her basket purse.

You can't answer the questions I have, I thought.

We shook hands again. I fought the urge to bow at the waist.

She stopped at the door and turned. "Oh, and I forgot to mention that they used to hang at the courthouse, but that probably won't help you," she said casually, as if she was giving me a recipe for whiskey sours.

"Thomasine mentioned hangings too," I muttered.

"Thomasine? Oh, the woman at the library?" Olivia asked. She wrinkled her lips and walked away. I followed her into the hall. As I watched her totter lightly down the stairs, I saw a body hanging, swaying, in front of the courthouse, facing the graveyard. Wouldn't the graveyard be the last thing the hanged would see before the bag went over their heads?

Olivia stopped. "Did you say something?" she asked up to me. I realized that I had spoken out loud.

"No," I said. "I realized I hadn't seen the graveyard yet."

"Oh," she said and continued out the front door. "It's right down the street!"

I heard the bell tinkle. For the briefest of moments,

I sensed what felt like a hand on my ribs, cupping the side of me, opening my waist like the side door of a house. I was aware of the separation of my ribs like I never had before, the spacing of them and their curves. I turned and was alone in the hall. Shaken, I staggered to the conference room and stared at Olivia's research, considering a prison break from the Center to the boneyard up the street.

Judy quietly entered the conference room. "Are you ready to meet with Delia?"

I jumped in my chair. "Yes. Yes, thank you," I replied.

Judy led me across the hall and introduced me to Delia, the executive director of the Dorchester County Arts Center.

"The ghost tour was originally Delia's idea," Judy said and left to work at her desk.

"I have opened up my mind to the possibility of phantoms," Delia announced. She's from Nebraska and has worked at the Center for eight years. She lives in Talbot County. She's tall and willowy with thick, white hair. She was wearing a simple and elegant pants suit. She's one of those people who talk in italics. Those people make the best storytellers. She told me her own personal High Street ghost story.

During her second year at the Center, Delia

was alone in her second floor office at the center and heard the bell on the downstairs front door ring. She called out; no response. She rose and discovered the door swinging open, so she closed and locked it.

"The minute I got back to my desk and my butt hit the chair, the bell rang again," she said with relish. She crossed to the landing just in time to see the door swing wide, despite the fact that she had locked it moments before. She closed and locked it a second time. The third time it happened, she said from the top of the stairs, "This is not funny. Stop."
It did.

According to my pagan friend Korinne, showing no fear and calmly asking a spirit to stop is the magic button. The voice of reason can chase the undead back to their haunts, and Delia knew this kernel instinctively. Ghosts know when we see them and know that we know that they know. It's a spiritual gentleman's agreement.

On a different and later afternoon in the same year, Delia was at her desk, alone again, and she heard the second floor supply closet door creak open. She made a creaking sound as she described the hook and latch supply door opening on its own. "Creeeeaaaak," she croaked. Again she found the door open, closed it, and returned to her desk. It creaked open on its own again. Again, she closed it and, again, it creaked open.

> *"I decided to ignore it," she said definitively, sweeping her hand across her desk stacked with folders thick with papers. She refused to close the closet door a third time.*
>
> *The creaking stopped, as if the ghosts needed attention and couldn't stand Delia's disdain.*

"You should narrate the tour," I said to her, taking notes. It struck me how detailed the personal, first-hand ghost story is in comparison to the vagueness of the handed-down tale. All the years of repetition out of a variety of mouths simplifies a saga. Or maybe ghost stories travel cloaked in parable so we can better understand them.

"Oh, no, not me," she said. "I won't do the tour."

"You're a storyteller," I said.

"I don't know about that. All I know is that I heard that door open a third time and I stood in that hall and I thought I've had enough," Delia announced.

I knew that feeling. I had stood in that hall and felt a ghost go by. I had stood in that hall and felt a hand cup my waist.

The hall lurked right outside my line of vision. I consciously tried not to sneak glimpses of it. I was not surprised that the area of spiritual activity that Delia experienced was the same hallway where the wall splashed my cheek. What was it about this space that created stories?

"We've heard people crying. We've heard walking on the steps when there's no one there," Delia said. "We hear cats when they are no cats."

Flowers' ghost category list doesn't include the spectral and neurotic call for attention. Maybe the Center's ghost is just lonely and bored for all of eternity.

An elder photographer, who uses the Center's communal dark room, brings a radio with him when he works so he won't hear the ghost's weeping. He's blotting out his sensory perception. Maybe most of the processing of sensory information is not about opening up but blocking out.

The Center building is several hundred years old, used to be a hotel and is soaked in transitory energy. I remembered the cold, wet sensation of something passing by me. I remembered the welt on my cheek. I remembered the feeling of being watched.

I told my friends Joan and Norrie about the eerie sensation of being watched at the Dorchester Arts Center. "What's the movie where they all watch? Oh, right, it's like *The Children of the Corn* or *The Stepford Wives*," said Joan, narrowing her eyes and tilting her head, thinking it through.

"It sounds more like *The Village of the Damned*," said Norrie. "All that denial. They want you to write about it but they have no ghosts." Norrie was born in Boston, and is a fine, smart writer with a very dry sense of humor and very dark lipstick. "It's like Freud's repeat repressed theory," she said. "They repress the same event over and over again."

But, why did Delia risk all the town push back for those two stories? Why did she wait seven years to tell them?

My red-haired musician friend Harp has read a lot of author Aldous Huxley's work and explained to me that our brains filter out the relentless waves of sensory information it receives from our eyes, ears, nose, tongue and skin. Huxley exposed this theory: that our minds sift through all that overwhelming sensory material and send only to our cognitive centers that which we can process, what we can handle. A door opening on its own, defying our basic concepts of physics and metaphysics, is too much for some folks to handle. Like the old man in the Dorchester County Arts dark room, we turn up the radio over the phantoms crying and work that way.

"We don't want to see behind the reality," said Harp as he played his bass on his sofa. He twanged a discordant guitar riff. "It's just too darn scary." He smiled and watched his freckled fingers as they played.

After Delia told me her ghost story, Judy returned to the office. Why had Judy left us alone? Did she think that re-telling the story was like a conjuring of spirit? Was she haunted by the stories as well?

Judy handed me a single sheet of paper with one paragraph typed on it. "Remember the request I made to all the High Street residents for ghost stories?" She asked me.

As I expected, her letters to the homeowners yielded little, one slight historical submission, knowledge that I had already gleaned from my thumbnail history of the County.

"I'm sorry that's all I got," Judy said wistfully.

"That's a shame," I said, at a loss for a civil reaction to the rich, white, Episcopal denial. "What can we . . ."

"Who might talk to Maryland?" Delia asked Judy. Delia and Judy reviewed a list of High Street residents to probe for ghost tales, ticking off possibilities. "We shouldn't ask Charlotte; her son committed suicide, and the Eveston's daughter was murdered in Baltimore," said Delia. "So steer clear of them."

This street holds its secrets, I thought.

Delia pierced me with her cool blue eyes. "Certainly, you can make up more stories or tell the folk stories from outside of town."

"I'm finding all the ghost stories outside of town."

"Well, you should use those. There are plenty of them, and you should stay away from what actually happened," she said. Judy stared at her hands folded on her pastel lap. Apparently, Delia thought that history's more powerful than folklore. I wondered the opposite.

I was short on High Street ghost stories, but most

of the county ghost stories are somehow associated with High Street residents. The legend of accused murderess Patty Cannon is tied to High Street through her legal counsel.

> *Patty Cannon was a 19th century bandit who kidnapped free African Americans and sold them back as slaves to their previous owners. She would sell kidnapped slaves to bootleggers, follow the bootleggers, kill them and re-sell the slaves, sometimes for $1,000 a person. Patty and her gang used her son-in-law's tavern in Reliance, Maryland on the Dorchester County line as a base for her nefarious crimes. A farmer plowing a cornfield near the tavern unearthed several skeletons, including several Cannon family members, and Patty and her gang were arrested. Cannon never went to trial because she killed herself in prison with poison that she had hidden in the hem of her skirt. High Street's Josiah Bayly, one of the thirteen Cambridge families and Maryland's first Attorney General, supplied Patty with her legal counsel.*

"I don't think we should talk about Patty Cannon," said Delia. "It's such a negative story about the Bayly family. Surely, you must be able to find other stories."

I bit my lip and said nothing. Josiah Bayly represented Patty Cannon over 170 years ago, at least six generations ago, the magic, native Cambridge number. Several Baylys have served as Christ Church vestrymen, there's a Bayly Street in Cambridge, and in the 1850s Christ Church parishioners were buried in the private Bayly graveyard adjacent to the church's graveyard.

Patty Cannon's ghoul still haunts the tavern in Reliance, Maryland. Tavern owners complain of eerie screams and loud footsteps. Stereos, TVs and fans turn themselves on and off. The she-devil Patty Cannon still lives on in story. PBS produced a Strange Mysteries documentary about her legend, but I can't talk about her on the tour because I might offend the surviving Baylys.

"But don't you worry," said Judy, patting my arm. "We'll find you other ghost stories."

She and Delia suggested that we visit Cambridge House, a charming bed and breakfast at #112 High Street and interview its proprietor. So Judy and I strolled down street to Cambridge House, a massive Victorian cottage with a wide veranda. William Vans Murray, Netherlands ambassador and exile from the Jacobite Revolution, was raised in this house. Judy and I teetered on its vast crooked porch. High Street houses are so old that most of its porches, like the Eastern Shore, are fighting to break away from their main foundations. Judy stood on tiptoe to ring the bell.

People this rich don't have ghosts, I thought.

After an extremely long pause, the proprietor, Stu, answered the door. I forgot to ask where he was born, but he's a Yankee, that's for sure. He was barrel-chested and sweating, and his skin had the yellowed tone of disease. He was practically dripping sweat. I wondered if I could resuscitate him if he collapsed from an attack. He invited us in for a quick first-floor tour, but he had no stories to tell.

"I'm sorry. We have no ghosts." Stu said. "I've tried, but this is a happy house. I was stuck in here alone, way before the renovations and my partner moved in, on a snowy January night, and I thought; well, now I'm bound to hear one, but nothing. I do have to say that I always had a warm feeling of children in here. I'd like to say that I hear children playing as their mother waited for her sea-faring husband to return to land, ha, ha, imagine that, but I've heard nothing. No little footsteps running up and down the stairs." He laughed and wiped his brow with a damp handkerchief. "I got sick not long after we moved in. Diagnosed, I should say. Been here seven years and these people still don't talk to me. Seven years and all I get is: good morning," he complained.

"Seven years," I muttered, staring at an overstuffed sofa detailed with dancing lords and ladies. Seven years ago, the same year of Delia's door stories.

Stu was selling the house because of health reasons; he was tired of traveling by ambulance to Washington. He felt trapped between the river and the creek, trapped like the typhoid victims. We stood on the broad porch after our brief tour. That far down the street I could taste the slight brine of the Chesapeake in the air. Stu gossiped, pointing up High Street, towards the Center.

"That house across the street from the Center was just bought by a lady from Atlanta. Now, she's Southern. She pulled up alongside me in a convertible and a big hat and trailing scarves like Isadora Duncan. When I told her I was leaving, she offered to buy Cambridge House, even though she had just put a bid on hers. Can you imagine the money? Good God almighty! But I already had a contract with those people from Pennsylvania. You should hear her; she sounds like she's from Tara." Apparently, Stu's not from Pennsylvania or the Deep South.

Judy had told me on the walk to Cambridge House that she was "originally from Charleston" which confused me since I thought she had said that she hailed from the Eastern Shore. In either case, she didn't like Stu's Deep South comment. I watched her upper lip twist as she turned to look north towards the river. A breeze rippled her short, strawberry hair.

Disappointed, Judy and I wandered to the curb. "I know one place on High Street that's chock full of the dead," said Judy, looking upstreet.

"The cemetery," I said and started off. Finally, I thought, the graveyard.

A curved, shoulder-high, brick wall contains Christ Church cemetery. We entered through the black wrought iron gate on the High Street side. The gate happily creaked, welcoming us, as Judy tugged it open. The grass was meticulously shorn but the ground rolled in uneven waves.

"Must be hard to mow," I muttered.

Two massive oaks shadowed the yard. In the humid afternoon, Judy and I searched for Ann Weller, the woman whose grave is engulfed by the hulking monster tree closest to the High Street wall, but we couldn't find her. The tree was hiding her too well; it had swallowed her.

The graveyard floor was damp and bumpy; I wore big, clunky boots to hike through it. I had survived knee surgery a few years previous, and my right knee felt particularly watery and wiggly in that graveyard, like it did in the scary weeks prior to the surgery. Maybe the water table full of dead people pulled at it like the tide. With every step I expected to sink ankle-deep into the soggy ground, bony fingers pulling at my bootlaces. Most Dorchester County burial vaults are shallow and not buried deeply because of regular tidal flooding. Diggers don't have to dig far to hit water here. Often the ground's too wet to bury, and internments are delayed because of flood.

A magnificent bit of flood folklore from the southern end of Dorchester County in Hoopersville, a town on the southern chain of Dorchester islands, tells the wild tale of a very wet internment right before the storm surge of Hurricane Hazel. Standing unevenly in the graveyard, I stared at the courthouse across the street and thought of the old fisherman.

> *An old fisherman had just buried his wife*
> *of many years in the family plot next to their*
> *house by a creek. As a fast and violent hurricane*
> *whipped the storm surge up over the family*
> *plot and to the steps of his groaning house, the*
> *grieving fisherman was convinced that his friends*
> *would come to rescue him, and, indeed, in middle*
> *of the howling wind, he heard a knocking at*
> *the door. He opened the door and there was his*
> *wife's coffin, floating on the rising floodwaters,*
> *its lid loose and the water holding her up in a*
> *seated position. The water fanned out her hair*
> *and lifted one arm out to the fisherman. She had*
> *found her way back home. Feeling the house rock*
> *on its foundation, the old fisherman grabbed onto*
> *the coffin as his home sunk under the waves, and*
> *he rode the flood in the coffin with his dead wife*
> *until morning when he landed on high ground*
> *several miles away.*[35]

In Flowers' version, the old man ditched his wife's body and rode the coffin alone. In the *Chesapeake Book of the Dead*, the old fisherman is an old woman.

We don't want to lose our dead. We want them to continue to guard and guide us, as if omnipotence is granted to us when we lose our bodies. I sat on a gravestone in the cemetery. A bird sang. The breeze ruffled my red skirt, and I missed my great-grandmother, Nannie, who died in my parent's house when I was fourteen.

> *Nannie had lived with us since I was*
> *two, and she died in her chair in her bedroom*
> *directly below mine. A week after Nannie*
> *died, I was doing homework in my bedroom.*
> *I was answering questions on* The Man in
> the Moon Marigolds *about the character of*
> *Nanny when I heard my Nannie call my name.*
> *I stopped writing and turned, but the warm*
> *yellow light of the room spilled only on my*
> *twin bed, my dolls and my first record player. I*
> *blinked and held my breath. She called again,*
> *clear, ringing and strong. It made me feel safe,*
> *as if she was letting me know that she had*
> *successfully crossed. She had reached the Other*
> *Side. She had reached home.*

I can still see the amber glow in the room and feel
the secure promise of her voice. Maybe I don't tell
my first-remembered ghost story because I loved that
first ghost. The memory still makes me sad. Some
thirty years later, and I still miss her. Korinne's right;
grandmothers cannot be in Purgatory. The branches
rustled in the tree above me, and a crow cawed. I found
myself staring at a crooked gravestone of Major Thomas
Nevett that read: *There is a gloomy vale between us. Pass*
through. I've gone before.

Many Eastern Shore homes have family plots, and
since the land is constantly flooded, it's not uncommon
to see submerged headstones in Dorchester County.

"Sometimes we open a grave," Christ Church's
Father Martin said, "and there's nothing there. Two

hundred years does its work." As he cackled at his own joke, I felt especially temporal. My remains won't exist in 2200. All traces of me will be melted into the breast of the earth by then, like Nannie's ashes into the dirt of Lutherville, like the Revolutionary dead into the Cambridge graveyard. It's scary and weirdly comforting, yet why do we turn to dust? Why can we not turn to something nobler, like light? Like the light that came for the Hill neighbor.

> *Mr. Travis from Aisquith Island in southern Dorchester County told me that he has several graves on his property and he once found a freshly decaying leg in a tomb that dated from the 1760s. Time and weather had cracked the masonry, and he could see inside the sepulcher.*

> *"How's that possible?" He railed. "I swear it was fresh! I could smell it! Nothing else smells like that! It couldn't be more than a couple days dead. In an open crypt from the 1760s!"*

> *"That's not physically possible," I said, writing notes madly.*
> *"I know!" He said.*

With heat and sans formaldehyde, flesh will rot within a week, expelling noxious hydrogen sulfide and methane. Within ten years, the skin has sagged off, and the body is toxic. After a couple hundred years in wet

ground, even the bones are gone. Living with the open
maws of your submerged ancestors makes it hard for
a people to repress their temporality. Maybe swamps
produce ghost stories as well as methane and weird
lights. Maybe stories match their topography, lush and
wet and reeking of death.

I wanted to wander though the Christ Church
graveyard, pulled randomly by each hole's bounty,
but Judy flagged the sextant to allow us access into
the church. She waved to me from the Church Street
graveyard gate.

"Come on now, Maryland, let's get a move on," she
called.

I felt a surprising surge of companionship with my
vaguely preppy guide. I felt strangely out of our time, as
if she was calling me from a hundred years ago, instead
of across the bumpy cemetery, as if I had once heard or
was to hear a voice call my name across that stretch of
memory-drenched earth. I followed her to the Church
Street entrance.

The current stone Christ Church is the third parish
church on that soggy spot; the second parish church
burnt to the ground on Thanksgiving Day in 1882.

"For a town this wet, there seems to be a lot of fires,"
I said to Judy as we climbed the steps, wondering if the
fires were somehow tied to the dead.

"Buildings used to be made of mostly wood then,"

she reminded me. "And people used open flame to see at night."

We entered the Christ Church chapel, and the cloying smell of centuries of the Anglican prayer books settled around us. Judy proudly showed me the rose-stained glass window and the communion kneelers, intricately embroidered with idyllic Eastern Shore scenes.

"It's a Tiffany," she said, nodding to the window from Britain. She clasped her hands behind her back. I got the oddest impression that she was crossing her fingers. "It's not signed, but we have a letter in the church office that says that it is one." The afternoon light threw rose shadows on the dark wood and cream plaster walls.

I fidgeted. I wanted to search for Ann Weller under the encroaching tree of myth.

"Next we go our appointment with Ellen the librarian," Judy said. This was the first I had heard of this meeting. Judy sounded so much like Mary Poppins that I almost expected her to clap her hands and say, "Spit, spot, children."[36] An image of her in a tattered, patched coat and a floppy floral hat, carrying a talking umbrella, flashed before me. I smiled.

On our way out of the sanctuary, Judy stopped in the church office and introduced me to the church secretary, an unflappable Native American, balanced behind a mammoth mahogany desk. Her internal peace provided a comfortable companion to the shelves of Anglican books.

"Iris is our Native American connection," said Judy. Iris smiled without opening her mouth.

Iris is a member of the Eastern Shore Tribal Council and works as the Episcopal Church secretary. A jar of store-bought cookies sat on her filing cabinet. She silently handed me a copy of the church's history, Great Choptank Parish. Its hunter green binding reminded me of my high school yearbook.

"Thanks," I said, taking the thin book. I opened its glossy pages randomly to an 18th century chart of the graveyard dimensions. In 1706, the courthouse plot was where the church is now, and the graveyard was one plot of land away from the corner of Church and High Streets. Apparently, buildings and bodies move on High Street. "The graveyard moved," I said.

"Yes," said Iris. "Graveyards move." I felt the brain exposure as I had experienced when the hard man in Fells Point stared at me.

"Iris," Judy asked, "Do you have any ghost stories to share?" She sounded like she was talking to a ten-year-old. Iris shook her silent head and her long, straight, cedar-tinted hair swung against the spotless, paper-free desk.

"What about the mysterious noises during vestry meetings?" I asked. "Olivia says that she's heard noises in here."

"Those noises we hear near the altar," Judy prompted loudly, like she was talking to the deaf.

Iris shrugged.

The spirit world's not surprising to her.

Her eyes were wide and green as new grass and full of walls. Her mouth was still.

Judy and I left for the library. As we crossed Church Street, an older woman in a blue-flowered print housedress and slippers limped by. She seemed out of place in high rent section of Cambridge. Before I could comment, Judy called, "Get along now!"

The Cambridge Library is built of solid, red brick on the site of an eighteenth century mansion named The Hill which is ironic since Dorchester County has no hills. The library's front porch arches look toward Gay Street, and its rectangle box side faces the graceful lines of the Italianate courthouse across Spring Street. Wedged between the parking lot and the northwest library corner, leaned two simple gravestones.

"Are those graves?" I asked.

"Those are the Woolfords. They used to own the house that was here."

"They seem awfully close to the parking lot," I said under my breath as I followed Judy. We entered a decidedly unimposing side door and walked up a narrow flight of steps to the main library floor. I could see the Maryland Room tucked into the corner.

Judy and I met with the head librarian, Ellen, and

her architectural historian husband, Winslow, in Ellen's cramped office. Ellen's big-boned, an intelligent yet guarded woman. Bored at a wireless connection meeting that morning, she drew a diagram of High Street on a napkin with fabricated ghost stories for each residence. She showed it to me proudly. "See? Lawyer ghosts, skipjack ghosts, servant ghosts, pet ghosts. You know, all the levels of society." She smiled. "Oh, no, you can keep it," she said, pushing it across the desk to me. Ellen has lived in Cambridge for eighteen years, and on her second New Year's Eve, she polled her fellow partygoers for ghost stories and came up dry.

"No ghosts here," she announced with Yankee conviction. Of course, nobody spilled; she's not a native. Nobody's going to talk to her. She's one of those people from Pennsylvania, even if she is the librarian. "You'll have to look outside of town for ghosts," Ellen definitely said, as if that decision had already been made.

Behind Ellen, a magnolia tree swayed in the arched window, its leaves turning backwards. Nannie always said that turning leaves was a sure sign of rain.

"That's where most of them seem to be," I said.

"It's very wet outside of town," Winston said. We stared at him blankly. "You hafta build up," he explained.

Ghost stories flourish out there, but nobody wanted to talk about them unless they happened safely out in the swamp to the lower classes. The bogs and marshes of

Dorchester County are full of mystery. Strange things happen there, so strange stories grow legs.

"Read this," Ellen said, stretching over stacks of books to hand me a sheet of paper.

> *A young man named Wallace walked twenty miles to his brother's funeral. Carrying his brother's coffin to the gravesite, he stumbled. Scottish folklore says that if you stumble at a funeral, the next death will be your own. Concerned by the stumble and saddened by his brother's death, Wallace drank heavily at the wake and left late to walk through the dark to his home. Part of his journey home took him through Dank Hollow, a swampy area where the road dipped down. One wrong step in the dark could land him in the bog and sink him "in the marsh that knows no bottom" as the Native Americans say, but the young man was resolute. He had told his wife he would return that evening, so he set out into the night. When his wife received him at their door at daybreak, he was delirious and fevered, and, despite a doctor's care, he slipped into a coma and died. On his deathbed, he whispered his frightening story to the pastor. As he reached Dank Hollow, he saw a line of shadowy figures approaching from the bog, one of which was the brother he had buried that day. They beckoned Wallace to join them. That was all he remembered, but when he collapsed at his door all his clothes were wet and turned inside out.[37]*

"Was that in Dorchester County?" I asked, aware that the others were watching me read.

"Oh, it was somewhere on the Eastern Shore," Ellen said.

"There's a Key Wallace Drive by Blackwater," offered Winston.

Ellen leaned across her oak desk and whispered, "I don't know what that last clothes-inside-out part means, but it's scary, isn't it?" Despite the broad expanse of desk stacked with papers and books between us, I cringed backwards further into my library chair until I pushed it against the wall behind me.

What scares me about this story is the image of a foggy line of undead moving across a bog, outlined against a row of ragged pine trees behind them. It's not possible for a human with any weight to walk straight across a bog. Marsh is scattered with islands of reeds in sickening lakes of black mud and brackish water. To transverse a fen, you have to jump between those slippery reed islands. A line of zombies steadily approaching across a bog might be your first clue that you've crossed to the Other Side. I couldn't get the image of it out of my head. I winced in my chair and bit my lip.

We fear the unknown ritual waiting for us at the moment of passing out of our bodies. A percolating swamp, persuasive with decay, seems an appropriate location for such a crossing. After all, man crawled out of the swamp millions of years ago.

We all hail from swamp, but it's still not possible to walk a linear line across one.

I offered the sheet of paper with the marsh tale to Judy. She shook her hand as she was declining a cup of tea.

"Is High Street built on Indian burial ground?" She suddenly asked. "Somewhere near the jail? I heard rumors of noises in the courthouse or at least a lot of busted pipes."

Winslow shifted his girth, and his barrel chair creaked. Ellen raised an eyebrow, and I sat up straight. Judy's question was the first time anyone had mentioned Indian burial ground, but I had researched Cambridge's connection to a 17th century Indian Reservation. A burial ground on that Reservation made sense.

In 1669, to appease Native American complaints about the loss of their land in the influx of English settlers, the Colonial Government granted the land on the south side of the Choptank, bound westerly by Sewell's Creek and three miles into the woods, as a Native American reservation. Included in that parcel was the current location of High Street.

In 1671, merchant John Kirk bought a nearby plantation named Ricarton.

In 1683, feeling threatened by the warring Five Nation Iroquois, the Choptank Indians took refuge with the neighboring Pocomokes and Assateagues.

In the Choptanks' absence, the Maryland government re-surveyed the land and incorporated part of that Indian Reservation into Ricarton.

In 1702, the Winacaco chief dealt with further English land encroachment by selling or leasing the western end of the Reservation to John Kirk, and that land became Cambridge. The chief sold 3,000 acres for forty-two match coats. A match coat is a cloak-like garment made of European fabric.

In 1787, the Confederation Congress of the United States passed an Ordinance that stated that the Indians' "lands and property shall never be taken from them without their consent; and, in their property, rights and liberty, they shall never be invaded or disturbed."

We all know how long that lasted.

Regardless of the complicated details of Maryland property exchange, High Street land was once a Native American Reservation.

"I thought John Kirk purchased the land from the Winacaco," I said.

"He owned Ricarton . . ." started Winslow.

"For a bunch of really nice coats . . ." I continued.

Ellen shook her head; her long gray hair struggled against its headband. "I doubt that any buildings were built over any graves," she said. "The new Cambridge Hyatt was built on the last of the Indian Reservation." The new Hyatt's down river several miles from High Street.

"When the courthouse was expanded, they excavated," started Winslow.

"We don't know that definitely . . ." interrupted Ellen.

"I had heard this about the courthouse, that they found something," said Judy, perching on the edge of her library chair, her eyes bright.

"Workers uncovered Indian remains and the Smithsonian carted them away," Winslow volunteered abruptly. His voice cracked on the word "carted."

I was thinking that the Confederate dead in the Christ Church cemetery across the street would never be carted off to the Smithsonian Museum when I felt an odd vibe in Ellen's office. Winslow had paled considerably and was wiping his moist hands on his jeans. The air smelled faintly of sweat and fear. Ellen loudly re-arranged a stack of books on her desk. I wondered if she was building a wall with the tomes.

"When did that happen?" I asked Winslow quietly.

"Probably in the later part of the 19th century, sometime in the 1880s," he said, stuttering on the date. He studied the carpet and picked lint from his pants leg.

Sometime in the 1880s, around the same time, the church burnt to the ground.

"What source do you have?" I asked.

Ellen changed the subject. "Don't you worry about that crazy story. That's just some old wives' tale. It's silly what Winslow believes," she added.

He's an architectural historian, I thought. How silly is that? Why did he tell it? Aren't we looking for ghosts? "Might be a good story," I suggested.

"There's an Indian Bone Road by Bucktown," Judy said brightly.

"The Native American Graves Protection and Repatriation Act of 1990 applies to any organization that possess Native American human remains and associated funerary objects and that receives federal funding," recited Ellen.

Did she memorize that? Keep it cribbed in her drawer? Her objections were all about a several hundred years old land fight, and the Indians were trapped in the middle.

Winslow rocked in his chair and gripped an architecture book. His wide fingers were pale.

"There are others. What about the Locust Street ghost?" Ellen asked. "Calvin used to haunt Pat but he stopped." Two blocks over from High Street, a spirit resident named Calvin haunted a live resident named Pat. Live Pat told dead Calvin to stop opening doors and turning on lights unless he started paying some utility bills, but dead Calvin refused. Not until his surviving wife died, did he depart the haunting business. I wondered if the old lady that Flowers wanted me to contact on Locust Street knew Calvin and Pat.

As I recount these ghost stories, I too am ridiculously assigning anthropomorphic qualities to flimsy energy, but that's part of the humanization of parable. We can only describe the unknown with behavior that is known to us. Maybe there are friendly, the Ghost and Mrs. Muir type ghosts, like the dead Calvin category that require only companionship or compassion. Dead or alive, we all want empathy.

"She died, and the lights stopped turning on by themselves. Maybe Calvin joined his wife; maybe he left town," speculated Ellen.

I doubt the existence of anything as structured as a town in any afterlife. I could be wrong.

"We don't want to offend," reminded Judy.

Risking offense to dead Calvin's decendants, I put his tall tale in the tour sans formal names.

"When did becoming a ghost become such a social offense?" I asked without thinking. "The advent of Christianity?"

Judy smiled. "That would do it," she said.

"And there's no need to mention the day in 1970 when they blew up the second-floor ladies room in the courthouse," Ellen announced in her office. Winslow nodded like a poppet, and Judy joined him. "It would be in bad taste," Ellen added.

"I read of no deaths related to the bombing," I said. "Or injuries to people. The building was injured."

"How right you are," Ellen said. "No ghost story to tell there."

Like the slave basement tale, I thought. I had to

ignore and pretend away history.

"There are more stories outside of town. Harriet Tubman lived south of town," Judy said.

"You can look up her court trial. It was right across the street in the courthouse," Winslow insisted, still wringing his doughy mitts. "It was in 1850. It was. I know the Clerk of Court." Harriet Tubman was a native of Dorchester County and was the courageous slave who developed the Underground Railroad. She strategically placed her safe houses outside Cambridge in remote Quaker pockets. I wondered if the Quakers believe in ghosts.

"Can you give me the phone number of the Clerk of Court?" I asked Winslow, still writing rapidly. The humid air of the afternoon settled around my slumped shoulders and smelled of rain. Clouds blocked out the sun.

"I don't know what help he'd be," Ellen said slowly.

"The Harriet Tubman Museum's across town," offered Judy.

I wanted to research the building of the jail and the possible Indian artifact removal. "The courthouse might have some ghost stories," I said. "Judy said she had heard about clanks on the pipes."

"We all have noisy pipes," Ellen said. "We all have flooded basements. It floods a lot here."

I didn't like it, but I wouldn't chronicle the bombing. Still, the courthouse seemed a good ghost source.

"Didn't they used to hang in the courthouse?" I asked.

"Oh, yes," said Winslow, wiping a lock of brown hair off his forehead.

Ellen leaned over her expansive desk, nudging the towers of books, and whispered, "They used to hang in the tower at the back of the courthouse, but in the 1880s, the public hangings moved inside." Her bottle green eyes sparkled. "But that was a long time ago. The only noises in there now are the old radiators."

Between public hangings and an Indian burial site, the courthouse was evolving into the best ghost story on the block. I had to get that court contact number.

"And Christ Church isn't the only burial spot on High Street. There are two graves right outside the library," Ellen continued, distracting me from the courthouse jail.

"Yeah, I saw them," I said, taking notes.

"They're unmarked and probably from its manor house days," she said, " but I think that's what happens to Baltimore writers when they don't return their library books in time."

Everybody but me in the room laughed.

"Ellen, you know that's the Woolfords!" Protested Judy, still giggling.

"Well," Ellen said, suddenly standing, "I have another meeting." She handed me a copy of Trish Gallagher's collection of Maryland ghost myths. "Read this."

Apparently, the interview had ended, and Judy and I were excused. Winslow shook my hand and scuttled away, head down, sweating.

"You should call journalist Brice Stump at *The Daily Times* since Tom Flowers won't talk to you," Ellen instructed. Stump has published several collections of Eastern Shore folklore, including a few ghost legends. "He might be able to tell you some stories, and if you can't find any, you can always make them up. That way they could all fit a common theme. Delia and I think that the tour should be about how the human spirit moves forward."

I was more interested in how stories evolve and why we tell the stories we do.

"What *Daily Times*?" I asked instead. "In what city?"

As we left the library, Judy carefully handed me Ellen's wrinkled napkin sketch, like an archaeological artifact. "You keep it," she said. She patted my arm absently again. I peered at it, but the handwriting was too small to decipher. I shoved it in my backpack.

As Judy and I walked through the intersection of Spring and High Streets, the same older woman in the housedress and slippers struggled by. I wondered if she might be a ghost.

"Do you see her?" I asked.

"The old lady?" Judy asked. "Oh, yes, I think her

name's Mildred. She walks up and down Church Street."

Maybe the old woman is the embodiment of how the human spirit moves forward: slowly plodding; spine bent; eyes down; ankles swollen, crossing the same intersection again and again.

Standing outside the Center at twilight, Judy apologized again for the lack of ghosts. "I wish I could hand you a list."

"Just get me the clerk of court number," I said.

"Ghosts or no ghosts, I have to say that I enjoyed the day," she said.

I was glad, so had I. ""Why did you ask about the burial ground?" I asked, trying to imagine the courthouse in the 1880s.

"Oh, it's just another rumor you hear. I must've heard Iris tell someone."

"What kinds of noises are in the courthouse?"

"Oh," Judy said, shrugging, "you know, just noises. Pipes. Ellen's right. It rains a lot here. All of our basements flood."

"Didn't the Arts Center used to be a hotel?" I asked.

She nodded. "Buildings hold energy."

"They sure do," I agreed.

Crickets played underfoot. "I almost forgot to give you the cuttings," Judy said. As she ducked into the Center, I noticed that the tomatoes along the porch were thriving; yet it seemed an early bloom period for tomatoes. Judy returned with a small, gray bucket half full of water and three tomato plants. The plants nodded over the lip of the bucket, suspended, sleeping, waiting. "I'll need the bucket back when you next come down," she said. I thanked her. I love gardening and knew exactly the spot where I would plant them. I doubted that tomatoes would transplant well after they had bloomed, but I accepted her gift. It was sweet.

Judy left me to fire up the Center's kiln for its pottery class, and I sat in the car, examining the napkin diagram. It seemed like the notes were taken in two different hands. One footnote that listed some book sources was in a decidedly different handwriting. Had someone at the wireless meeting helped Ellen plan the ghost walk?

I exited the car and walked the path of the tour, drawing my own map. The diagram in the Cambridge History Tour brochure didn't match the information in the *Cambridge Past* book, a collection of the town's photographs with several awry High Street lot numberings. The town can't agree on house numbers, much less stories. So I mapped the street myself, pacing the path of the tour.

The sky threw large droplets of occasional water on the notebook page, and the paper wrinkled like a pirate's cartography. My map looked like a child's drawing; my simple lines didn't convey the majesty of the architecture or the echo of the voices under the wind. I felt watched by the houses. My knees popped as I strolled. I passed a historic marker that notated the formation of Cambridge in 1684.

"When the Indians were outta town," I said to no one in particular.

Harriet Tubman's code song of *Swing Low, Sweet Chariot* rolled through my mind. When a slave heard that song in the quarters on a pre-Civil War Saturday night, he or she knew that a Railroad guide was not far.

Maryland's first political split was not the Civil War; during the American Revolution, Chesapeake watermen sided with the loyalist Tories because of their hatred for the rebel Maryland gentry. Some watermen traded illegally with British warships for handsome profits; some looted American ships.

A culture's ghost stories paint its history in rich dots, like scattered rain on dry earth.

Across Long Wharf and the Choptank River in Talbot County, locals tell a Revolutionary War legend of a headless man, a British sympathizer waiting on the beach to guide British vessels to an attack on St. Michaels. St. Michaels is a charming town up the Bay from Cambridge that was spared a Revolutionary War battle because someone decapitated the Tory traitor on the sand. Forty years later, the British attacked St. Michaels during The War of 1812.

> *The headless ghost spy strides back and*
> *forth across the Talbot sand, scanning the dark*
> *horizon, holding a spyglass up to where his eyes*
> *should be. Waves crash and wind blows away*
> *the phantom's footprints. On a lane not far from*
> *the same beach, the same ghost has been seen*
> *astride a spirit horse, madly galloping along the*
> *winding road and narrowly missing overhanging*
> *tree branches, even with the extra low clearance*
> *that headlessness allows.*

This Talbot County tale has a ring of Promethean repeat. What "fresh hell"[38] is a nightly, headless ride for all eternity? How did the rider lose his head? Like Big Liz, the horseman has a double death whammy in the decapitative part of the story. A headless ghost is twice as scary as a headed one; headlessness is death archetype times two.

Dorchester County's ghost tally was climbing: two headless stories, two dead wife at the door stories and two vanishing stories.

At the end of High Street, the road opens into a roundabout with the American Cruise Lines dock to the south and the public marina to the north. By the water stands a memorial smokestack from one of Franklin Roosevelt's ships. I stood by the grounded smokestack and photographed the river expanse to the Talbot County side, the headless horseman territory. Maybe I could catch the ghost spy on film. The clouds scuttled across a gun barrel day. The seagulls swooped and yelled. The wind filled my eyes with water and painted a rippling darkness on the river surface. I blinked away tears. An amorphous black shape zipped along the stone retaining wall. I rubbed my eyes.

To the right or east, over by the entrance to Cambridge Creek, two men in a rowboat traversed the choppy transition where river meets creek. Both men were older and wore dark clothing; one wore a slouched hat. They moved gently, as if they suffered arthritis or were trapped in a state of suspended animation. One man slowly rowed the boat while the second carefully, rhythmically, methodically reached into a bucket and threw a wriggling fish back into the river. The fish leapt in perfect arcs back into the water, catching the remains of the day on their scales. It seemed almost like a dance. It almost seemed like ritual. Why would anyone throw perfectly decent fish back into the river? I remembered the dream I had the night I first visited the Dorchester Arts Center, the dream of the men in a line of boats, throwing something back into the water.

Joseph Campbell theorized that mythology began with early man's recounting of the hunt, describing how the killed animals appeared to transition in death to a spiritual world. Early hunters created the concept of an animal master who sent the animals back to earth, to be sacrificed again and again in the endless cycle of death and food. Hunters believed that they had to appease the animal master or the prey would not be returned and the hunter's family would starve.[39] That appeasement became sacrifice and soon took on the pattern and import of ritual. That was the beginning of religion, and the creation myth circle of birth, life, death, resurrection and ascension still flourishes.

Boating is dangerous on the Chesapeake Bay; the currents are complicated and fierce. Maybe Indians make a sacrifice each time they push off; I know I used to make a little prayer each time I sailed. Indians believe that spirits inhabit water, along with corn, sun, thunder, lighting, fire, animals and stones. Maybe a couple of fish could appease the temperamental river god.

We make up patterns in stories because we're scared. We need ritual patterns, like we need stories.

I finished my map and limped back to the car.

May 30th

PORCH PUDDLES

The Eastern Shore gets very dark at night, and I was afraid to look in the back seat of my car as I drove home. My eyes flicked to the rearview, and I remembered an Easton ghost story my friend Shannon had told me. Shannon was born on the Eastern Shore and is a certified actor combatant with soft auburn hair. She and her sister see things.

> *A year ago, Shannon's sister was driving along Route 50 near Easton around midnight, and she saw a young man in a plaid shirt with a hand in his pocket and a cigarette in his mouth. A mile passed and she saw him again. A mile later she saw him a third time. It's physically impossible. It's* The Hitchhiker *come to life or to death a few miles north of Cambridge.*

I remembered her unnerving story as I sped away from the Arts Center, attempting to stay with the pack of cars as we rattled along the black envelope of Route 50. As hard as I tried to keep pace with the other cars, I kept finding myself suddenly alone with the nearest headlight half a mile ahead along the flat track of the road vibrating through the reaching corn. Afraid to look to the drainage ditch beside me, afraid I might glimpse the hitchhiker and see quickly into the Other Side, I hit the gas and caught up with a tractor-trailer. Its powerful pocket of pulsing energy rumbled beside my car, pulling it down the highway, along a linear path of time. The car wheels spun over the road, the tilting earth spun around the sun, and the moon spun around the earth. The radio tied me to the ground as Sarah McLachlan sang *Drawn to the Rhythm*. "In the face of a blinding sun, awake only to find that heaven is a stranger place than the one I left behind."[40] I compulsively looked in the rearview for a dead man in the back seat with a lit cigarette.

All these thoughts of temporality, all this consideration of the conversion out of the body, made me grip the wheel until my fingers ached. One fast turn and the transition out of this relative certainty would begin. The truck door shimmered beside me, a haunting gate to the Other Side. I'm curious to know but terrified to experience it. I'll pass through the eternal mystery soon enough, but, once on the Other Side, I'll probably not recognize consciousness the way I do now. Looking through an adjacent car's window at the rippling fields was like a quick glimpse into the Other Side: everything's slightly warped, tilted, and a little different than this world. One expects to see a ghoul. Maybe the Other Side surrounds us constantly on a molecular level; there's lots of empty space in molecules. For all I know, the Other Side could be right next me daily.

Despite my hitchhiker fear, I de-accelerated and let the truck pass; its lights sped towards the horizon and the Bay. The bucket rich with tomato plants sloshed softly in the back seat, and I swayed in this reality a little longer. I felt outside myself, the foreigner, as if I was watching my life in a film, beyond it and inside it simultaneously. When my car clacked over the last span of the Bay Bridge and I was safely back on Maryland's Western Shore, I breathed a sigh of relief and felt oddly decisive.

Tell the stories, I thought. Like a returned gift, like throwing the fish back to the Hunter God.

I was beginning to suspect that the writing of a story results in a new version of truth and I was transitioning into a new version. At the very least, the ghoul stories from out in the boondocks were creeping closer to Cambridge, trailing their carrion fiction behind them.

Did the dead care what story is told? I wondered. But that would assign consciousness to phantoms.

By the time I reached home in Baltimore, the earth had turned far into the other side of night, so I left the tomato bucket on my back porch. As I set it down, I found four tomato plants inside. I thought Judy had given me three but the fourth must've been hiding, tucked beside the others.

"This new reality must have more tomato plants," I speculated.

A small, fast black shape, darker than my garden's night, whirred along the edge of the porch railing. I shook my head and went to bed.

I had a quick early morning dream that I was debarking from a long boat that docked by the edge of the Christ Church High Street graveyard gate. Iris, the Native American secretary, was curled in a tree, offering me her hand, full of crumbled cookies.

"What's the river doing over by the church?" I asked her.

She shrugged. "Graveyards move," she said in a deep man's voice.

I thought I woke up, burrowing deeper into my flannel sheets. Something exhaled very sharply on Karl's side of the bed.

I'm not alone in the house, I thought in a panic. I felt pinned by the blankets. I heard the sound of footsteps at the bedroom door. I felt the bed frame begin to lift and jumped up, the sheet tangling my ankle. Heart pumping, I searched the house.

"I heard two short, distinct "ha has," I said, describing the phantom panting moment to Korinne on the phone. "The room was blindingly bright, and Karl was not there. Maybe I was asleep, but it sounded so real."

"You should put up the wall during this project," said Korinne. "Surround the house with light." Even with Karl in bed beside me, before I fall asleep, I visualize a pulsating wall of illumination engulfing the house, from rafters to basement. The concentration exhausts me and I pass out. Karl can attest that I can fall asleep mid sentence. I try to remember to imagine the light each night, like prayers, but sometimes I forget.

"You can't forget," said Korinne. She sounded so serious, and I hadn't even told her about the black shapes. "It's too dangerous," she said.

"Why?" I didn't want to know.

"I can't say," she finally replied.

Does this happen to other folklorists? I wondered. Or only ghost folklorists? Vernon Griffin's online biography said that he had died. I wondered how. I wondered what he knew now. Folklorist George Carey's biography stated that he had retired to Maine to raise bees. Was tending bees safer than folklore? I kept thinking about Flowers as he was transported through the Green Briar Swamp.

I called Brice Stump at *The Daily Times* and left a rambling message. "I just want to tell the stories," I blathered. I fought the urge to ask if he saw black shapes or heard phantom breath or felt a hand at his waist.

Under a threatening sky, I poured the Cambridge water from the tomato bucket into the garden by the

side basement wall. The water seemed to percolate
as it entered the soil, sending up tiny clumps of dirt,
bubbling into the ground. The spring had been wet and
so was the dirt, and the tomato water was deliberate as
it sank. Vaguely creeped out, I dug a swift hole, and
still the earth walls leaked tobacco-colored fluid from
the surrounding garden. Wet roots seemed to reach
for my gloved hands, so I planted quickly, shaking off
determined tendrils. After I replaced the dirt around
the tomatoes, I buried my face briefly into their feathery
leaves, and the sharp, tangy smell of tomato leaf filled
my head. I rinsed the bucket with the outdoor spigot,
and as the residue tomato water swirled into the house
foundation, the pipes in my house shuddered and
clanked.

I wonder if the courthouse clanks resembled those
clanks.

I know an electrician named Shawn, a native
Baltimorean who's tall, redheaded and thoughtful. As an
electrician, he's heard all sorts of odd house noises.

> *"We were upgrading the electric in Notre
> Dame College, in a dorm where the old nuns
> used to live. Do they call that a rectory?" He
> asked. He scratched the damp kerchief on his
> head.*
> *"Anyway, I had to run some cable through
> the attic. Electricians knock on pipes to
> communicate when we don't have walkie-talkies,*

*but my partner was downstairs on the third floor
and nowhere near any attic pipes. So, I'm up in
this attic by myself and the pipes start banging
like crazy on the other side of the building. I
followed the noise and it moved with me, always
countering to the other side of this round attic.*

*I went downstairs and asked my partner if
he had been banging any pipes and he said no.*

*'We're done with this job today,' I said.
And we left."*

Shawn laughed when he told me this story, his belly
shaking under his coveralls but his blue eyes were serious.
He wasn't about to stay up in that attic with any dead,
pipe-banging nuns.

Pipes thump daily in my house, but it's an old house.

The next day, I checked on the tomato transplants
and somehow they seemed impossibly taller. Thinking
of Shawn's ghost nuns, I climbed into the wet garden
and leaned against the side of the house, listening for
knocks. My garden clogs sank in the damp, loose dirt.
My cell phone rang.

It was Judy, checking on my progress with folklorist
Brice Stump.

"Thanks for the tomatoes. I planted them yesterday
in the side yard," I said. "I left Brice a message."

"I hope he calls you back," Judy said. "Ellen said he
would."

"Maybe she could call him," I suggested. I wished that I had Judy's faith in the shaky Cambridge ghost story network.

"What I really need is a contact at the courthouse," I said.

She sighed. "I'll do what I can," she said forlornly.

"Even with Delia's stories and the graveyard and the stuff from outside town, I'm a little short on story," I said. "I don't have enough for every house on High Street. What do you think about the break idea?" I had suggested that the tour skip several houses in a row before the wharf.

"Skip some houses? Can't you make something up?" She asked. "Can't you find another story and put it in there?"

I bet the we're-too-rich-to-have-ghosts homeowners might object, I thought. "Couldn't the tourists just walk for a while?" I asked. "And roll over what we told them?"

God forbid Americans ruminate in silence, I thought.

"What about all those stories that Ellen gave you?" Judy asked.

A pipe in the wall by the tomatoes knocked. "I suppose so," I said. "People need time to process ghosts," I said. I didn't say that everyone in Cambridge was reacting like most of humanity to the ghost stories riddled through the thin fabric of their damp community. We ended our phone conversation, and I went inside to read more ghost stories.

I spliced one Maryland phantom tale into a wavering hole in the High Street tour, when I ran dry of the elusive unicorn that is the Dorchester tall tale. My source for many Maryland ghost stories is Trish Gallagher's *Ghosts and Haunted Houses of Maryland*, a wondrous tome that's chock full of levitating candles, strange whirlwinds, mirrored phantoms and swinging chandeliers. It's certainly a book worth reading, but none of the stories took place on High Street in Cambridge. Frenchtown's a tiny hamlet in Somerset County on the southern Eastern Shore, and Mt. Airy lies west of Baltimore, halfway to West Virginia.

> *In a Frenchtown tavern, a young girl walking upstairs saw a disembodied hand on the staircase. The hand grabbed her ankle, and she fell forward, screaming. When live family members appeared, the hand vanished, but the girl's ankle was bruised and swollen.* [41]

That story's just plain creepy. The staircase could be a big metaphor for the march of time, and the hand seems to be holding the girl back in the past. Do ghosts anchor us back to the past? Do their stories?

> *An old lady living alone in a Mt. Airy mansion didn't like anyone using her front parlor and resolutely refused to use it herself. She kept*

> *it meticulously clean. She died from a fall down*
> *some stairs. The night before the funeral, her*
> *relatives laid her body out in the front parlor*
> *since it was the best room in the house. The*
> *morning of the funeral, the parlor door was*
> *locked and its key could not be found. The*
> *relatives broke into the parlor, and they found*
> *the parlor door key, lying next to the old lady's*
> *casket.* [42]

I like that old lady story, although I doubt that the dead have the same housekeeping concerns as the living. I admit that I plugged the tale into the High Street tour. I altered folklore by moving the guts of one story somewhere else, to settle into High Street, like an old lady struggling into a new coat, like Mildred crossing Church Street. It galled me that I changed the phantom traditions, yet people amend folklore over the years. By the pure nature of oral tradition, the storyteller alters some detail in the tale each time. This is how legend evolves. A few years of ghost tours and people will believe that puddles appear on High Street porches and the parlor door locks itself in #110.

I cannot have been the first to alter folklore. People change it daily.

Gallagher's stories seem close to a plausible reality. Most of her stories, except the wild hand legend, have a core of simple, credible truth. The pages are thick with benign slamming doors, aimless footsteps and wispy phantom smells of cigars and apple blossom perfume. The doors in my house regularly slam and open without my help. Pipes bang and radios turn themselves on. Not all ghost folklore is random; some have seemingly causal consequences. The ghoul stories with consequences scare me the most. One from Gallagher's compilation is a believable, unfulfilled task message from the Oaklands in Laurel south of Baltimore.

> *A woman in a long Colonial dress appeared on the lawn and beckoned a male resident to the edge of the woods. Confused and scared, he followed her. She looked alive, like she could be an actress in period clothes. When he got a few feet from her, she pointed to the ground. He looked down into the ground, and in the tall grass, the man found a gold locket from the 18th century. When he looked back up, the phantom had vanished.*

> *Since that first sighting, the man has awoken, paralyzed under his sheets, hearing approaching footsteps and eventually feeling a heavy breath over him, immobilizing him.* [43]

What did the man do with the old necklace? Did he ever research it? Did he visit all the local pawnshops? Pay someone for a title search the land? Sell the mansion? In some way, he seemed like he didn't want to know. It's very much like how we all wrangle with the question of our own bloodied individual pasts. Where do we store it all? We must do something with it or it will continue to breathe hotly on us as we sleep.

> *I've been immobilized by phantom breath, and I have experienced nothing more terrifying. When I lived alone in an old row house in the Mount Vernon section of Baltimore, I would sometimes awake, knowing I was not alone in my bedroom. Something very powerful was nearby; I could hear it breathe. I could smell its decay. It could smell my fear. I felt a presence behind my head, leaning over me, freezing my muscles, hot breath on my forehead, an ancient tongue rattling through my ears. Eventually, I found the strength within me to fight back and move my leaden arms to turn on the light and read some passages from the New Testament book of St. John to calm my terror.*

Was it only a dream? How could it be anything but a dream? Why did I continue to dream it? And, when I moved from that apartment, why didn't the dream follow me to my new home until these recent, odd, breathing dreams seven years later?

Seven years again.

> *Since the 1770s, members of a small community south of Cambridge, called Craddock Marsh, not too far from DeCoursey Bridge and Blackwater, swear they have heard an eerie breathing, crying and howling of a mysterious half-man, half-beast they call the*

Yahoo (yay-hoe) Monster. It gets its name from the eerie sound it makes, and one resident says, "at times you could hear it, both day and night, for several weeks."

One Craddock Marsh resident claims a sighting of the Yahoo Monster. He had been hunting in the marsh, and the sun was setting. He was collecting his gear and preparing to head home when he heard a splashing and smelled what he described as the fierce smell of skunk that permeated the area. He swore he saw a pair of red eyes flashing and more water splashing noise before he ran back to his truck.

Still, others who hunt and trap in the marsh have found no tracks of anything but the usual swamp animals: ducks, squirrels, deer, muskrats and turtles. One hunter who experienced the monster said "he was scared because the noise was so strange and it seemed like it was all around him all at once."[44] The sound was particularly terrifying because it completely surrounded him; its source was not just one direction.

That's the sound you never want to hear: the sound of the hunter becoming the hunted.

The mysterious Yahoo surround sound has a *Blair Witch Project* tinge to it, that all-persuasive, enveloping noise with no obvious source or end. How can people live with it? And, yet, people have resided in Craddock Marsh for over two hundred years.

How different is their suppressed terror from the terror of the unknown that whips around us all every moment of every day? How do we temporarily repress the awful knowledge of temporality? French absurdist playwright Eugene Ionesco believed that the only true absurdity was that we all know that we are all going to die. We just don't know when or how. That's the only true suspense of our lives, the cliffhanger mystery of our lives. We have all experienced the spiritual or other-worldly. Maybe those stories can help us unravel the mystery. Western cultures don't have an easy acceptance of mystery in the world. We want to solve everything. We should embrace the mystery. It's out there.

My friend Casey claims that she has a ghost in her apartment that laughs at her from time to time, when she drops something or does something ridiculous. Casey hails from Kentucky and is mostly logical. She walks with a bounce in her step and has a collection of adorable hats. Despite her outward calm, when Casey believes in something, she believes in it fervently and she believes that her apartment building's haunted. Years ago, Casey had a strikingly real and very mysterious experience that confirmed her belief in ghosts.

*Casey was working in a theatre in
Bloomington, Indiana one night years ago when
she and a whole team of technicians saw a
woman in a long coat over a longer dress walk*

up a set of stairs to a locked door and disappear.

"She was in the whole Victorian thing, you know," Casey said. "The dress, the coat. She was even carrying a muff and everything."

It seemed logical to the techs that the woman might be an actress in costume; she made noise and seemed alive. But the fact remained that the woman vanished into a locked stairwell through a locked steel door and that denies physics and questions metaphysics.

"She seemed so alive," Casey said. "You know, she had weight, like a person."

The Victorian lady had weight like the man in the Travis hunting lodge that the boy mistook for a living man. How can we possibly distinguish between the living and the dead when the dead can so easily masquerade as one of us? How can we possibly interpret the oblique message they deliver? Do they deliver a message or are we projecting the hope of one? Where is the Rosetta Stone for the Other Side?

The day after I planted the tomatoes, I had a slight fever and fell asleep on the couch with the television still on. I woke up to the sound of my dead grandmother, calling down to me from my upstairs bedroom, her voice clear and strong as when she was alive.

"Maryland!" She called. "That girl's getting arrested!"

I looked at the TV and the movie Rear Window *was playing. Jimmy Stewart was frantically telling his policeman friend over the phone that Grace Kelly was getting arrested.*

I sat up and went quickly through the stages of ghost denial. At first I was a little freaked out, because my grandmother's voice sounded so true, so like her. Then I started missing her. Then I got angry that she hadn't sent me a better message, some clearer clue to the mystifying universe.

"I wanted more," I said over the phone to Korinne. "And not, basically, wake up and watch the good stuff!"

"Yeah, well, but," Korinne wisely pointed out, "that's not a bad life message."

This morning, there were two puddles on my back porch, but it didn't rain last night. The rest of my street was dry, the porch chairs were dry, and the lawn was dry. I looked for an oyster shell but there was none.

Serves me right, I thought. That'll teach me to make up porch puddles stories.

Maybe the old lady in the parlor story was not as far fetched as I assumed it was. Maybe mysterious events could be literal and linear. Maybe I should be more afraid of stairs and hands on them.

Maybe writing is a type of conjuring.

If only I could conjure a Cambridge folklorist.

June 3rd

BAD LUCK BLUE BOATS

I called Father Martin, the previous rector of Christ
Church, to arrange an interview. Judy couldn't find
his phone number, so I scored his contact information
through mutual Episcopal friends. I had heard gossip
from two sources that he was asked to leave his rectorship
over a music program dispute.

"Have you talked to Olivia?" He barked his first
question to me, as if she would explain all.

"Yes," I said. "We met a couple weeks ago. She said
there weren't any ghosts in the graveyard."

"Of course," he replied with firm assurance.
"Ghosts aren't real."

We scheduled a phone call in a few days.

I fought the urge to ask him if the Tiffany window
was real. I wondered if Judy had crossed her fingers
behind her back when she mentioned it. Was it bad luck
to talk about the Tiffany window?

Dorchester County clings to a myriad of rules and protections against a variety of life's complications. A rat leaving a ship spells immediate disaster for the vessel. Women are not only banned from fishing boats, but to meet a woman on your way to work is bad luck (which is a clear indication of the small size of the county's population). People aren't encouraged to sew on Sunday, sing at the table or in bed, spill salt, rock an empty rocker or start housecleaning on Friday. A falling picture is a sure sign of death. Painting a boat blue is bad luck.

I remembered that the boats in my dredging dream were painted blue.

We might scoff at these superstitions but modern culture still follows many of them. We still walk around ladders. We throw salt over our shoulders. Professional athletes practice superstitious behavior, which becomes self-fulfilling prophecies for success. Superstitions, like legend and myth, change and grow. Ancient Egyptians believed that the space under a ladder was to be avoided because it was a holy triangle of the gods. Early Christians replaced that holy triangle with the Holy Trinity. In the Middle Ages, the fear of being doused with boiling oil during a fortress siege added another potent reason not to walk under ladders.

Daunted by the staggering power of nature, early man connected gods to lightning, flood and earthquake. Early Greek philosopher Pythagorus theorized that earthquakes happened when too many dead gathered underground. Superstitions grew from man wrestling with an understanding of his universe, and myth and religion developed in a parallel fashion, from a similar base. The etymology of the word religion has roots in the verbs *to connect* or *to link*. If we know we are going to die, we want to know why. We want to feel like our life has had value, worth and reason.

Each of my writing projects defines my reason to be alive.

"Quick question," Judy said over the phone to me. She wanted some tour stories to feature in Dorchester Arts Center marketing flyers.

"I'll send you what I have," I said.

"Can you finish writing the tour by the end of June?" She asked.

"There's an eighty percent possibility," I said slowly, carefully. "I haven't found a narrative voice or finished all the research."

I heard myself tell excuse stories to Judy. *We tell ourselves stories in order to live.* Did she believe mine? They were true to me. I wanted to ask her about the church window.

"Why's the research taking so long?" She asked.

I should've been interviewing more Cambridge residents, like a folklore journalist, like the way I was dredging ghost lore out of my hesitant friends, but I didn't have the time or the money to travel to Dorchester County and no one was talking anyway. "There was lots to read and research, and no one's talking," I explained, trying not to whine.

Judy sighed sadly. "I know," she admitted. "It's tight here."

No one wanted to talk ghost.

My Cambridge sources were cagey, and I didn't completely trust them. They all, with the exception of Stu (and he's leaving), seemed to have some sort of block in place. They were guarding something. They were hiding something. Were they evasive because I'm the outsider, a foreigner in my own state? Or were they hiding something that's too big to talk about? Something about Indian burial grounds?

When I listed my sources to Thomasine, the pirate-denying volunteer from the Maryland Room, she said, "Don't worry, you're talking to the right people."

"Well, that's heartening," I muttered, "but they tell me little."

These repressed people were my option for oral, non-published folklore, and the reliability of all human sources is dubious at best.

Cambridge natives aren't the only ones who don't like to talk ghost. Most people don't; the subject makes them uncomfortable. I'm not alone in my experience of Eastern Shore ghost denial, but I didn't tell Judy that over the phone or I might've sounded like I was bashing the Eastern Shore.

My friend Sid was involved in staging a mystery tour in Easton, and his researchers received the same "no ghost" denial that Cambridge gave me.

"The Eastern Shore's screwed up," Sid said. "Surrounded by all that swamp and in complete denial of death and ghosts." His people hail from Winchester, Virginia. He's a quiet carpenter; he's seen life. He squinted at the clouds and wiped a line of sweat from his high forehead. "The South is definitely not over the Civil War, and the Eastern Shore's definitely Southern."

Did Cambridge clam up because they thought me Yankee? Sid's Southern, but the Eastern Shore wouldn't talk to him either. Maybe he's not the right kind of Southern. Maryland and Virginia have been arguing for centuries, mostly about water rights and oyster beds. Maybe the Eastern Shore folk don't want to reveal their ghosts to someone from western Virgina.

"But Cambridge has ghosts! I know all about ghosts in Cambridge!" said Sid proudly, smiling. "Denial or not!"

> Sid was on honeymoon years ago in an isolated house on the Choptank; a friend had let the couple use her home. Isolated houses in the middle of swamp are common south of Cambridge. From the moment Sid and his new wife arrived, they heard footsteps and doors closing at night. On daylight investigation, Sid discovered a locked door at the top of the house. His wife said the owner told her that she never

used that section of the house; it was too big.
The next day, after more footsteps at night, the
locked door was open.

"When the footsteps started up again and
the television turned on by itself, we holed up in
the bedroom with the cat," said Sid, grinning at
the sky. "That cat was so terrified it crapped
all over the floor." Finally, after three days and
nights of torment, the newlyweds called the
owner. The owner admitted to housing a spirit.
He occupied the shut-off part of the house.

"Oh, I didn't think he'd come out with
strangers," the owner said, as if the phantom
was a socially impolite guest.
Sid and his new wife left immediately.

What was natural to the owner was supernatural
to Sid. The newlyweds had the common sense that
the *Poltergeist* and *Amityville* families lacked; they left the
haunted house. They heard that deep voice resonate
in their hearts and heads: *Get out. Get out now.* They
listened and they left. Of course, Sid didn't own the
property, so it was financially easier for him to bolt than
the *Poltergeist* family.

Ghost stories shut up a lot of people. I was recently
at a party of theatrical artists. We were sitting around
a bonfire in a late spring Baltimore backyard, drinking
beer. A small hole opened up in the conversation,

punctuated only by the sputter of the burning wood.

"We should tell ghost stories," I said. It seemed the perfect ghost-telling atmosphere: a fire, beer, and the open night sky. There was an awkward silence.

"You know plenty," said my friend Tony. Tony's a native Baltimorean. He makes fine bean soup and can manage to smoke while he rakes leaves. He's the person on the planet that probably understands my brain the best. He introduced me to the poetry of Charles Bukowski.

I began the story of the tour and this book. The actors looked worried; some stared at the ground. A newcomer arrived and interrupted the tale. The actors laughed too quickly. The fire crackled.

"Ghost stories always stop the conversation," I said, taking a swig of beer.

Tony smiled quietly beside me. "I didn't think that would fly," he said into his half empty bottle.

"Why are we so afraid?" I whined.

"There's no place to put them, the stories. We have no place to store them, no room in our belief systems," he said.

If you think of your belief system like a set of shelves, where do you stick ghosts? There's nowhere to put the headless swamp girl and the hand coming up through the stairs. Where can they fit beside daily events or big, bulky beliefs like: religion, the non-linear nature of time, temporality, and the hundreds of thousands of

people who have died before us? What happened to
their energy?

But I couldn't say that to Judy over the phone.
Instead we discussed the voice of the tour.

Judy was a little worried; she was in the process of
auditioning tour guides and didn't have many candidates.
She wanted to use both genders as guides to double her
limited possibilities, but I wanted to place the narrative
voice solidly in either one gender or the other. "The tour
guide has to have a gender. It's a character," I tried to
explain.

The guide's voice will embody the message of the
tour: live life and watch the good stuff. I wanted to
use a character from a ghost story or a historical figure
to convey that theme, but I also wanted a resident of
Christ Church graveyard. I was torn between Major
Thomas, the Reverend Maynider's drowned son-in-law,
or Willimina, the woman who abandoned her true love
at her father's bidding. They both slept in the Christ
Church graveyard. Major Thomas Nevett was the dead
guy with the fabulous epitaph: *There is a gloomy vale
between us. Pass through. I've gone before.* He'd be the perfect
guide, reassuring us all that something, even a gloomy
something, lies beyond the tomb.

I liked the idea of him being drowned and
conducting the tour soaking wet. "You know," I said to
Judy. "He'd be dripping."

Judy didn't think that a drenched guide was a logical idea. "How would he get wet and stay wet?" She was probably right.

"Maybe just his hair could be wet. It'll be summer, right? Something about a guide dripping water creeps me out," I explained, "And I could connect him through his family to Hannah and her grave robber story."

"Wet guide or dry guide. Voice or no voice," Judy said firmly, " I'll tell Delia that you'll deliver the finished tour by July. We have to keep Delia and the board happy." Suddenly, sweet substitute teacher Judy had transformed into a headless version of an editor. I didn't like it, but clearly I had to follow their deadlines. "Oh, and the name of the tour might change," Judy said. "I'm talking to the board. I'll let you know."

"That's fine," I said.

The Dorchester Arts Center might own the specific High Street tour, but they can't own the stories. No one owns them or everyone owns them. Society shares them. The stories have no definite authorship; that's part of the definition of myth and legend.

"Don't worry, baby, it's not possible to defame the dead," said Karl. He knew this from his Stalin and World War II research. He and I were curled up under the sheets. "You can say anything you want about dead people."

"I just want to tell the ghost stories," I said.

"Might be bad luck. Do you believe in ghosts?" he asked me. We had been dating a year and a half, and maybe we should've discussed this big belief topic sooner.

"I think our perception of them is a lot like our grasp of religion. We know something's there, but we can only give it a story that we can understand, so I'm assuming that that story is way off. What's true or real if anything is. When the best and brightest of us only use twelve percent of our brains on a good day, how can we possibly hope to understand a world without our bodies? How can we interpret one hundred percent of reality with twelve percent of our brains? There has to be more out there than we are capable of perceiving."

He rolled over on his back, staring up at the sheet tent above us. "That's a pretty good answer." I rubbed my nose against his shoulder.

"How can man presume to know god?" I asked.

"Hubris," he said, "the Greeks' first big sin."

"It'll bring us down." I traced my finger on his upper arm. "I've seen some weird things," I added. "A lot more than I care to recall."

I believe in energy, in the movement of energy through the world. Energy moves and its vestigial shadow lives in trees, rivers, cars, boats and houses; but every time I made a sweeping new age statement like that, Karl mocked my Feng Shui habits.

"Baby, you think the position of your chairs changes

the world," he said.

"Why is that so ludicrous?" I said. "The chairs were once trees." I told him the story of the haunted tree in Henry's Crossroads where the pine trees tower over the telephone poles. Henry's Crossroads is barely a town now; just a few houses and a stop sign in eastern Dorchester County.

> One night in 1910, walking in a wooded path by his home, the Sheriff of Dorchester County, Harry Turner, saw a woman in a white gown standing beside an ancient and towering pine. She stared at him and then suddenly climbed the tree with an astounding pace and agility. Turner watched, transfixed, and, as the phantom woman reached the top of the sixty foot tree, she gradually faded away into the mist and disappeared. Startled, Turner began walking the road armed, and twice had the opportunity to shoot the specter. Bullets flew through her as she climbed the tall pine only to vanish at the top. One Crossroads resident named Davenport encountered the same ghoul on the same road at night, but, instead of climbing the pine tree, she continued to grow bigger and bigger until she was the size of the tree and then vanished into mist once more.

> Some ghost stories are braided with older remnants of pagan spirits; the Henry's Crossroads story seems to be more about a tree spirit than a female ghost. Folklorist Brice Stump reports that the Henry Crossroads

> *community couldn't explain the phantom. Some*
> *connected the pine ghost to the mysterious holes*
> *that appeared in Julian Horseman's backfields*
> *and the midnight screams that rang through*
> *Hornickey's Lane around the same time.* [45]

"There is mystery," I said.

"Were they drunk? Did they have fear of women?" Karl asked, solidly rooted in reality.

"There are more things in heaven and earth, Horatio, than ever dreamt in our philosophy,"[46] I said, quoting Shakespeare from 400 years ago.

"We have no idea who really wrote that," he said, referring to the still-debated Shakespeare authorship issue.

"Well, no. We don't. But we also don't know who started the pine ghost story or Big Liz or Hannah Maynider's stories."

"People did. The same people who wrote the Bible did. I'm going online," Karl announced, throwing back the covers.

I felt dread for some reason and researched prophetic Dorchester County dreams to the sound of his machine-gun keystrokes. "I have to listen and open up my ears to the possibilities of the universe," I said to the empty bed. Karl didn't respond.

I'm glad that there are more things in heaven than what my limited brain can imagine. I need to trust what brain I use and trust its intuition. I have

inherently known some random things in my life, and
if I've learned anything as I age, it's to recognize the
illogical knowledge of something. I knew when I drove
home from Cambridge that the stories would gather
in a book. Sometimes I can tell if a storm's coming
days in advance. I knew that Margot and Lynn were
Episcopalians. The morning of the day in 1984 when
I was robbed in Syracuse, I irrationally wanted to wear
my pearl necklace to work, but didn't, and thieves stole
the necklace from my house. When I was twelve, I woke
up, running down the upstairs hall, calling, "Nannie
fell!" Moments before, as I slept, my grandmother had
lost her grip on her invalid mother, and Nannie had slid
to the floor. I don't know how I sensed that they were in
danger while I slept, but I did.

Dorchester County has its second sight lore, its
legends of forewarnings, mostly of death. There linger
several stories of doors and windows that flew open at
the moment of death, blown by chilled air. There are
even more dream legends.

> *A man went duck hunting in winter and
> disappeared. As dangerous as the marsh is
> in summer, it's twice as hazardous covered
> in slippery ice. His sister-in-law repeatedly
> dreamt that her brother-in-law's boat was
> covered in blood and trapped in the ice. Finally
> she convinced her husband to investigate, and
> they found the brother-in-law's blood-spattered
> boat by the duck blind where she had seen it in*

her dream. They didn't find her brother-in-
law's body until spring, after it had frozen and
defrosted in the marsh for several months.[47]

That story isn't that far-fetched.

Dreams are considered prophetic. For thousands
of years, people believed that dreams were messages
from the gods. People have made predictions based
on dreams since before written language. Apocalyptic
prophecy describes destructive weather clearing the
path for the end of days, followed by a new world order.
Would the gathering of the dead underground bring us
living a renaissance?

French prophet Nostradamus described his dark
prophetic gift as an "emotional tendency handed
down to me from a line of ancestors." Sounds like the
definition of myth to me. History, like its first cousin
myth, is laced with prophecy.

According to historians Plutarch and
Suetonius, Roman dictator Julius Caesar had
several prophetic dreams. As praetor of Spain,
he dreamt that he was raping his mother. A
soothsayer interpreted Caesar's nightmare as
his destiny to rule the world, to plunder our
universal mother.

The night before his murder, Caesar's
wife Calphurnia dreamt that his statue spouted
gallons of blood.

Modern culture could use soothsayers again, interpreting the mystery that is dream. The modern version could be called archetypal psychologists, and they could sort through our sixth sense. Dreams aren't always just random bits of our memories; some dreams have messages. Premonition dreams abound. Several presidents suffered prophecy and had the strength to admit it.

> *Abraham Lincoln dreamt of being on a ship, headed towards shore, on the night before all of the big Civil War milestones: Fort Sumter, Bull Run, Gettysburg, and Appomattox. Not long before his death (some say three days before and some say the night before), Lincoln dreamt that he awoke to the sound of women weeping. He followed the sound down the hall of the White House to the East Room where he was lying in state after being killed by an assassin. In the same terrible month of April 1865, Mary Todd Lincoln told a confidante that she knew that her husband would soon die a sudden and violent death. Years prior in 1837, in Illinois, before she met Lincoln, she told a friend that she would marry a man who would become the president of the United States.*

> *In 1880, the night before President James Garfield was elected, he had a dream premonition that he would be killed in office. In July 1881, he was shot and died two months later.*

We believe our sixth sense in questions of love, and we trust instinct in romance. Why don't we believe it in the rest of life? Or is it coincidence? Is it luck? Can intuition allow us sight into the future through the folded blanket of time?

My friend Joseph is a high school student at Baltimore School for the Arts. He's a sound technician with a beautiful singing voice, a sweet disposition and a lumbering walk. When he was a child, he had premonitions. In one of his first ones, he was in church and looked across the aisle at a man. The man looked fine, but Joseph thought that man is about to fall. As he watched the man suddenly gripped his arm and fell over.

"Was it a stroke?" I asked.

"I don't know but I was so scared that I almost fainted," said Joseph.

How did Joseph know? Did he see the man's falling in the frame of time to come? Is déjà vu like looking into a different frame of time? What is déjà vu but prophecy and what is prophecy but a peek inside a parallel universe? Could the falling possibly be the powerful result of causal thought? A glitch in the program running the world? We have all experienced déjà vu, and we all discount it and ignore it. We shake it off like a chill.

I was teaching a theatre high school residency, and one of the students, Elaine, suddenly announced a déjà vu at the top of class. She designed the costumes for the school production of Julius Caesar. "And you were all in it," she said to the small group of us. She tugged at the flowered kerchief that held back her unruly black hair.

"It's nice to know that there's a parallel reality populated with all of us," I said, smiling." It's vaguely reassuring."

"It was all of you, standing around, before class, like now, and then Frank walked in." We all followed the turn of her head, and her fellow student Frank walked into the auditorium.

"Oooh," cooed several students together.

"That's whacked," said one.

"Don't trip on me," Elaine said. "It's not that weird." She back-pedaled away from her spiritual gift.

The students shook off their creeps and changed the subject.

"Well, good for you, Elaine," I said. "You admitted your second sight."

"Everybody gets déjà vu," she muttered.

I wondered why I should have two realities with Frank. One reality with him was plenty; Frank was stuck in a particularly obnoxious state of his development at the time. Maybe deja vu is as simple as seeing into another reality, a parallel one where Frank's a decent human being. Regardless, I was proud of Elaine because she bravely talked about the unknown. She pointed out déjà vu.

As I age, I try to train myself to listen to seemingly illogical instinct; it has skirted me around trouble in the past. But society doesn't condone prescience, as it used to, certainly as it did in Caesar's time. Psychics are often accused of over-exaggeration and hyperbole and the fraudulent ones give the others with the gift a bad name. I don't quite believe the sham spiritualists who claim to channel the voices of the dead, mostly because I can't imagine that intent or consciousness exists in the beyond. Ghosts are probably traces of energy, not conscious beings. I can't believe that phantom communication is that literal, but that disbelief doesn't mean that second sight or ghosts don't exist.

I have to believe in my version of the gift because I live with it and can't control it. We are shown what we are shown. It's another brain use issue that we don't know how to direct so we disqualify it. We can program artificial limbs to respond to thought command and we can land on the moon, but we can't cure the common cold and we don't use eighty-eight percent of our brains.

Suddenly and illogically, I felt an urge to check the tomato transplants. I stopped reading about second sight and was surprised to find an hour had passed. I staggered outside into the sun. When I planted the tomatoes, the four plants were about two feet away from the wall in a gap between two azaleas. Three days later, I swear they have somehow inched closer to the crumbling foundation by about a foot. They leaned away from the sun and towards the wall. They were reaching for the house.

I sat beside them and shuddered in the heat. Honeysuckle vines invade my yard but none of them grew anywhere near those Cambridge tomatoes.

Karl walked around the side of the house. "There you are," he said. "I have to go. I have an early day tomorrow." He had that pinched look on his face. His turning around to leave me was somehow scarier than the mutant Cambridge tomatoes or a second reality with obnoxious Frank.

June 6th

WORM WINDOWS

I heard the breathing in my bed again this morning
when I was alone. Karl was researching at The National
Archives. I thought I was awake and was lying in bed,
waking up, when I heard the scalding breath break the
morning light dulled by a thin rain.

"Ha, ha!"

Hot breath singed my right arm as it lay outside the
covers. I sat up abruptly and rubbed absently. Nothing.
I had forgotten to surround the house with light the night
before.

What was I thinking?

I checked my arm for damages. The skin wasn't
burnt, but the hairs were all flattened as if they were
wet. I brushed them up with a finger. I lit a candle and
opened the drapes, letting in what little sunlight was
swimming doggedly through the morning downfall. I
heard the sound of water rattling in the walls when
no water was running and the heat wasn't on. Still, I

feared a busted radiator pipe, so I searched the house for puddles. I was feeling around the radiator in the living room when I remembered that at 4 A.M. I had awoken to hear fully orchestrated Irish music that sounded as if it was being performed in the center of the intersection of streets behind my bedroom window. I sat on the rug and stared at the bottom of the couch. Maybe I could hear parallel realities, the other reality where my blue-collar neighbors play Celtic music in the middle of the night where Catalpha meets Echo Valley. Time's a wet blanket; it can fold on itself and double back. Or maybe I dreamt song from Ireland.

I don't want to believe in the breath.

My friend Terri's father, Andrew, was born in Brooklyn and occasionally smells cigarette smoke in his smoke-free house where the walls have been sealed against previous cigarette smoke damage. Andrew doesn't want to acknowledge the smoke smell. The previous homeowner was a heavy smoker who died in the house. Sometimes, late at night, Andrew hears phantom footsteps. "If I believe in the smoke and the footsteps, I believe in ghosts. And if I believe in ghosts, then I have to believe in God," he explained to his daughter.

"But what if ghosts are just leftover pieces of energy, random shreds of smoke?" Terri asked him.

I can accept the concept of random strings of leftover energy blowing by like smoke, like perfume, and

I think Terri's close to a simplistic understanding. I think physics has something to do with spirituality. The closer we get to observing the smallest of particles, the closer we get to an understanding of the universe.

As science developed machines that looked further into the atom, physicists realized that particles don't follow the old rules of Newtonian physics, and quantum physics evolved to describe how particles follow probabilities of motion and position. Newton's system paints a rather dismal picture of the Great Machine of the universe that was set into motion in the beginning and is not affected by the will of man. In Newton's world, man's impotent, and there's no free will. In the study of quantum physics, one can only concentrate on one aspect of an atom's address at a time, either motion or momentum or location. Since we can only know one probable description, we don't focus on the others, so we must project those probabilities, and therefore, we invent a different universe.

Either man has no free will or we create the whole shebang.[48] We're fools or gods.

We're a planet full of creators, making both time and reality. We produce money, presidents, deities, books, paintings, houses, radios, iPods, hairpins and ghost stories. As deeply held cultural beliefs, ghosts and their stories are human constructs like poetry, music and math. Our thought projections, like probabilities, construct reality, and we don't fully know the force of those projections. Meditation is energy; prayer has power. People fashion archetypes; we carve them out of shared experience and generations beyond us sense them.

I kept thinking about Flowers being transported through the Green Briar swamp and my friend John falling through a worm hole outside the haunted Glen Burnie house. Self-determination creates parallel realities; each new decision creates a new reality split. Flowers' decision to explore the marsh resulted in the new reality of a story. Wouldn't all storytelling do the same, every novel, every play, and all folklore? Is there a parallel reality where Big Liz is real? Does writing equal projection equal conjuring?

According to physicist Albert Einstein, this agreed-upon reality contains three spatial dimensions (up/down, right/left and front/back) and a fourth dimension of time. Modern string theory of particle physics speculates the existence of six additional dimensions beyond those four. Atoms are composed of quarks but no one really knows what comprises quarks. In 1984, two physicists theorized that tiny, unseen, electromagnetic strings are the fundamental building blocks of quarks and that additional, minute dimensions can exist at the curls of those strings. Unfortunately, strings are not testable because they're too small to be seen by current technology, so string theory has been in a bit of rut for several decades. Still, it makes sense to me.

The speed of light is a constant, and space and time are the flexible, fluid ingredients of Einstein's relativity equation. Both space and time are changed by the mass of objects that occupy them. One can apply Einstein's mind-blowing theories to ghost stories; they could explain how spirits seem to appear and disappear, to melt into the air and be made the wind. Time's bent around the big pine tree in Henry's Crossroads and the swamps of Craddock Marsh.

In 1915, Einstein developed this experiment while he was a patent clerk: he placed a watch in a vehicle that could approximate the speed of light. A second watch remained on the sidelines of the track. The watches were synchronized. After a few laps at the speed of light, the moving watch was slower than the fixed one. Maybe Blackbeard's moving at the speed of light, and the teenage boys are living at the rate of speed of the watch on the sidelines.

The more we discover about the universe, the more mysterious and complicated it becomes. There's speculation recently that known carbon-based chemicals only comprise ten percent of our universe and that a mysterious dark energy that we cannot see or measure is the rest.

> *My friend Casey's mother was driving
> home one day and suddenly knew that something
> was very wrong with her close friend. As she*

> *turned into her driveway, she briefly saw the*
> *friend sitting on her porch at the exact moment*
> *that the friend died in a car accident across*
> *town.*

Did Casey's Mom see into a parallel world? Some ghost stories sound like reality jumps, windows or worm holes into a different reality: the Colonial woman with the necklace, the Victorian woman Casey saw, Blackbeard appearing over the dunes, the headless spy, Big Liz, and Casey's Mom's dead friend. A worm window is one possible explanation of hauntings, and I don't want to discount any possibility. After all, I only use twelve percent of my very human brain.

I remembered a moment in my life that seemed like a time jump, as insane as that sounds. It had slipped into my consciousness like the memory of a long ago dream, a gleaming pearl buried in the shifting beach of my mind. It's my best out-of-this-realm-of-reality story by far.

> *My college friend Dave and I were*
> *preparing for a party at his house. Dave's a*
> *Baltimorean, and is usually very laid back*
> *but that night he had worked himself into a*
> *nervous frenzy, all concerned about the myriad*
> *of cookout details. I tried to calm him. He*
> *abruptly decided to change clothes. He bolted*
> *up the living room stairs, stopped midway and*
> *turned to tell me to answer the door if anyone*
> *rang. At the moment he turned, I saw Dave*

as a woman on a curving Colonial staircase,
and he was wearing a flowing pink gown and
a white wig. He was telling me to answer
the door if any of the guests arrived early. I
blinked. Dave blinked. The light seemed very
bright. We look startled. The 20th century
Dave raced upstairs. Weeks later, drunk with
him on tequila, I confessed my mad vision.
Dave admitted that he had turned around
and seen me, as a man, in Colonial garb,
at the bottom of huge staircase in a marble
entranceway. Dave was telling me to answer the
door in case any guest arrived early.

"Oh my God," said Dave, his head in his
hands, "What are we going to do?"

Well, we continued our lives and our friendship and
we lived, but what does one do with that experience?
Parallel reality or past life and either one with a gender
jump; it's a bit much to process. I had blocked out that
wild story and put it safely beneath my everyday realities.
It happened, yet I don't know what to do with it. Every
one of us carries stories about some mysterious spiritual
experience, and most of us suppress them. The thought
of them would force us to shift our bulky belief systems
and that's far too much work and far too scary. If we
allowed ourselves to see into other realities daily, we'd
lose our minds, or what this society defines as losing
our minds. We'd never get anything done. The gift
of sight can be a curse. We all possess the ability to see

into other realities if we only could allow ourselves to see. We won't open our eyes. We won't allow our brains to process the odd, ponderous baggage that makes no sense. Like Huxley's theories that apply to the old man photographer at the Dorchester Arts Center, we turn up the radio to block out the unknown.

But, Einstein was right; time is flexible. The *TV Guide* is a lie; time doesn't run in a straight line like that.

> *My pragmatic Minnesotan cousin Peggy was at a reception at the American Embassy in Oslo, Norway. She was chaperoning her daughter's dance group. Towards the end of the lunch, Peggy went to the ladies room and left it in what seemed like a few minutes to her.*

> *"It was super fast," she said. "No more than two to three minutes, tops."*

> *When she exited the bathroom, the whole group of twenty dancers had disappeared. The Embassy was completely empty and locked down, and a servant had to let her out. The dancers and Peggy's daughters had all vanished in a few minutes. She had to travel via bus and train to catch up with the group hours later.*

"The aliens got her," giggled my Aunt Doris when Peggy told me this story.

"She lost some time," nodded Casey when I told her this story.

Peggy lost about an hour and a half. She has no idea how. She wasn't sick; she didn't have a heart attack or a seizure. Maybe she walked back into the parallel reality where it was an hour and a half later. Maybe the group of dancers was moving at the speed of light like the Einstein watch experiment.

Casey's family moved from Kentucky to Indiana when she was in high school, and despite the move, she still lived about an hour and a half away from the Kings Island amusement park, only on the other side of it in Indiana. She and some Indiana high school friends went to Kings Island and rode the pirate's ship ride that rocks back and forth. During the ride, Casey took a photo of her new friends on the opposite side of the boat. When the photos developed, right behind her high school friends, was the entire family of Kentucky neighbors from Casey's childhood. Casey had not seen them on the ride or in the park, yet there they were in the photo: two kids, parents and grandparents, all as large as life and screeching with glee.

"My camera photographed another reality," Casey said.

Parallel realities have been mathematically proven. Can we slip beside a window and unknowingly see into them? How startling is that? What if we land in the wrong one?

In this shaky reality where I was commissioned to write ghost stories for a ghostless town, I interviewed Reverend Martin. He has a voice made for television. Despite mutual Episcopal minister friendships, Reverend Martin seemed to be guarding something in his conversation with me, testing me as much as I was questioning him. He had no ghosts to report, no weird vestry sounds or worm holes in the several-century-old Christ Church and its bumpy graveyard.

"It's a peaceful place. The organist, his name was Robert. I asked him how he felt when he was alone in there. Organists spend lots of time alone in their church," he said, pausing for dramatic effect. "Robert said the church was very peaceful." I wondered how Robert was involved in the rumored music program dispute. Still, Father Martin's comment rang an archetypal bell in my head.

> When I was working at the Shakespeare Theatre at the Folger, I lived with the scenic artist, Ruth. Ruth's from Connecticut. She has lots of thick hair like me but she's about a foot shorter. She's a fine painter with a husky voice.
> At the time, the Folger Library was the only combination library, theatre and mausoleum in the country; Henry Clay Folger's buried behind the reading room fireplace. One night, Ruth was alone in the theatre, painting, and saw a seat in first row fold down on its own and stay down, as if a weight was placed on it. She joked that

Henry Clay Folger had visited her.

"Weren't you afraid?" I asked.
"No. I got a very peaceful feeling, but
I have to admit that I started packing up my
paints not long after it happened." She tipped
back her delicate head and laughed.

Circumventing the peaceful comment, I tried to steer Father Martin towards some potential ghost stories, so we chatted about the Civil War.

"Lots of Confederate dead in the graveyard," I said, pointing out the obvious.

"The Eastern Shore was clearly Confederate," he replied with conviction.

"I know. It supplied gold and guns to Richmond."

"Maryland was a slave-owning state. Listen, don't believe that slaves in the basement story," he warned. "It's crap."

I frowned as I took notes. Did these people have secret, block-the-foreigner meetings? Did they hold those meetings in the church basement?

"What about Harriet Tubman and her Railroad stops?" I asked, fulfilling my promise to the Arts Council committee by focusing on other slave folklore.

Martin said that Harriet's owner was named Brodess, and he was a Christ Church member who died in 1849 and willed Harriet her freedom if she could claim it within ten days of his passing. The all male, all white Dorchester Court didn't inform Harriet of her inheritance, but when she heard rumors, she raised $5, walked the six miles through field and marsh from Bucktown and hired someone to read the will to her. "Furious that her freedom was dashed, she ran to Pennsylvania," said Father Martin.

I wrote wildly, wondering if he really just said, "dashed."

"She was a mystic!" He trilled.

> As a slave, Harriet suffered a severe head injury that almost killed her, and for the rest of her life she occasionally would fall into narcoleptic seizures that pushed her into a trance-like state. She claimed to hear the voice of God while in those trances and He would direct her along her dangerous path. During her Railroad trips, she claimed to follow an invisible pillar of cloud by day and a pillar of fire by night.
>
> Bucktown locals claim that the ghosts of Harriet and her parents, Ben Ross and Harriet Green, haunt the corn fields south of Cambridge, singing Swing Low Sweet Chariot, less than a half mile of thicket from the Green Briar Swamp.

Or maybe the farmers south of town just plain like that song.

I doubted that Harriet Tubman was on some phantom task quest. According to her folklore, Harriet left nothing unresolved in her time; she died quietly and was buried in Auburn, New York. Her tombstone reads: *well done, thou good, and faithful servant.* Her life was a freedom epic with a satisfying ending, literally a made-for-TV movie. Maybe people still want her as a guide to their freedoms, so the singing ghost stories bloom in the fields south of town.

In 1995, Father Martin officially received the approval of the Great Choptank Parish to add Harriet onto its Episcopal calendar of saints, as the first woman and the first black woman on the all-white, all-male, 150-saint roster.

"It shook up High Street when I got Harriet on that roster, I tell you!" Father Martin said loudly. He sounded like he was projecting his voice to the back of the church.

"I'm sure," I agreed, holding the receiver away from my ear. If the rumors were true and Martin was asked to leave the parish, I understood why.

"Have you tried talking with these people?" He barked, and then his voice softened and he took another sudden tack. "Have you talked to Tom Flowers? You should, you know."

"Tom's too busy with his garden to talk to me," I said sadly.

"You call him again and you tell Dr. Flowers that Father Martin said that he was a very important man to talk to." He was cracking himself up.

I didn't know what to say that was civil or Christian. "Well," I started, appeasing, "I feel sure both you and Dr. Flowers are equally important."

He cut me off. "Listen, for God's sake, don't say that the church has a Tiffany window. That West End tour calls the windows Tiffany and they aren't." Like Judy, the congregation claims that one of the church's stain glass windows is a Tiffany. During his rectorship, Father Martin had the window investigated and proved that it was not. The congregation was not pleased. They would rather cling to their no-ghost, no-slave myth, like a stained glass coat of colored light, protecting them from the storm of the supernatural around them.

Maybe that's why Judy crossed her fingers when she talked about the window.

I asked Father Martin about the hangings in the courthouse.

"The courthouse! Slaves were auctioned at Spring Valley," he crowed.

"Where's Spring Valley?" I asked. Winslow had mentioned that criminals were executed there; it sounded outside of town.

"It's across from the church, that little plot of grass on the side of the courthouse."

Spring Valley is about thirty feet square of bloody ground lying between the courthouse and the graveyard. A fountain of an armless woman now sits in Spring

Valley and, towards the creek, is a brick square with a memorial to John F. Kennedy. I was sitting on all possibilities of ghost on High Street, and nobody was talking.

"At the center of Cambridge," the Reverend said, "the courthouse meted out justice, the jail punished injustice and the church forgave sins." He cackled again. My skin crawled over my bones. I felt suddenly cold in 80-degree weather. "In the midst of life, we are in death. Do you know that one? It's part of the Episcopal burial service."

"One-stop shopping" as my friend Harp said when I told him this story later. He twanged a guitar chord. "I might write a song about it." He smiled. "Spring Valley. The 7-11 for all your spiritual needs."

Father Martin cut short the interview and said, "Well, if old Tom won't talk, then call Chief Winter Fox. He's the head of the tribal council, you know. He's good people. I have his number around here somewhere." The "good people" description was the most sincere comment he had made, the only time in our conversation when he didn't sound like a freaking game show host.

Finding no ghosts at Christ Church, I left Good People Chief Winter Fox a voice message. I tried not to sound too white. I can't help it. I sound white.

Sighing, I hung up and called Winslow to clear up

some High Street architectural discrepancies. The West End Tour pamphlet didn't match old Cambridge maps.

"Don't say that #110 is Victorian. It's not," he said.

"Right."

"And the courthouse is Italianate. Do you know how to spell that?"

"Yup, and speaking of spelling. About Harriet Tubman. I've found several spellings of her owner's last name: Brodess or Brodas."

"It's Brodess," he said. He spelled it.

Pick your history, I thought.

"So, did Harriet really raise five bucks, walk six miles from Bucktown, and hire someone to read the will?" I asked. I pictured her gathering the courage to walk up the courthouse lawn to the big, burnished front doors, rooted in bloodstained Spring Valley.

"It wasn't that straightforward," Winston said wistfully.

None of Cambridge or its history seemed straightforward to me.

Winston said that Brodess willed Harriet and several other slaves their freedom with the stipulation that they be freed at a certain age. By the time the will was read in a hearing at the Dorchester County Courthouse in 1854, Harriet had already escaped and was untraceable. That was the Tubman court record case Winslow mentioned at the library meeting, and I couldn't find any record of it.

I asked again for a courthouse contact number.

"It's in my other office," he evaded.

"I suspect that displaced Indian bones would shake up some energy," I said, carefully. Winslow didn't respond. "Or maybe I'm wrong." I left that sentence dangling.

"I have to go," Winslow said. "I have a meeting."

Judy mailed me a clipping from the *Dorchester Star* that detailed a recent Cambridge haunting. I found it odd that she mailed it with no phone or e-mail warning. Maybe she mailed it from the reality where she doesn't have a scanner.

> *Some people claim to have heard a dead man with a wooden leg clomping up and down the stairs of the Cambridge Market Center. An old sailor who lived above the Market Center had lost a leg in a fishing injury and had a wooden peg leg. Weeks after his death, vendors in the Market Center still heard his wooden leg banging up and down the stairs. Natives speculated in the article that the old sailor might be searching for his lost leg.*[49]

Maybe the Market Center stall owners were hearing into the past and not a ghost in the present. Which spectral quantum physics position is more probable? Listening to the past or being haunted in the present?

The Market Center is several blocks south of High

Street, but I considered putting the dead sailor story into the tour for a home near the graveyard that once housed a series of sea captains. I kept it as a back up plan if I should run short of High Street yarns. Would moving a story two blocks bend the fabric of space and time?

Everything does.

I e-mailed Jean, the peg-leg story's author and *Dorchester Star's* editor, for any other supernatural tales.

She wrote back: "Sorry. I've placed notices in the newspaper, asking for ghost stories in the past around Halloween with no luck. Personally, I have thought that it would be interesting to give a ghost tour in Cambridge but I never had the time to do the research."

I felt a little bit better about my outsider relationship with the blue book, clammed-up, ex-penal colony that is Cambridge. It was no use to interview the town folk; they won't talk to Jean either. I wondered where Jean was born. I wondered who spilled the peg leg story. I wondered if the old sailor had spare legs.

Dorchester natives believe that people should collect all of their body parts so they can be buried completely in death. A person buried incompletely is forced to roam, gathering missing pieces until all the body pieces are reunited. Some locals still collect their teeth for their eventual burials, and early Eastern Shore doctors saved discarded body parts to be buried later with their owners. Many Civil War ghost stories recount dead soldiers returning to the battlefield hospitals to retrieve lost limbs.

> *One southern county folk story tells the tale of a longhaired woman who was suddenly deathly ill with blinding headaches. No doctor or home remedy could revive her until her grandmother remembered an ancient warning to "Never lose any part of you."[50] Her longhaired granddaughter had a habit of emptying her hairbrush into the backyard. The relatives searched and found a deserted bird's nest laced with the sick woman's long hair, stuck in a dying tree. When they burned the nest, the woman fully recovered.*

After I read this story, I compulsively checked all the trees on my property. I have a thick head of long hair, and big clumps jump off my head and decorate the bathroom floor. I didn't find any bird nests, but one twirling white strand was woven through the mutant tomato plants. I pulled it slowly away, and the stalks bent towards me and sprung back to the house.

Feeling a headache coming on, I took two aspirin.

I vowed to check the plants daily. I considered ripping them out of the ground, but, then, what would I do with the bodies? Weren't they carrying Cambridge energy in their green cells? Seemed like simple physics to me. What energy were they pulling from my hair?

June 7th

HUMAN HERE

I awoke this morning to blinding sunlight. The night before I had remembered to surround the house with brightness. I didn't have the breathing dream. I dreamt that I was spinning, and Cambridge whirled around me in a stone church and clapboard house blur. Big Liz whipped by in the buggy with the Talbot doctor. I reached out to hold onto something, anything, to stop it, but I only felt water. Then I woke up.

When I put on my glasses, I saw a faint water stain on my cream bedroom wall that wasn't there the night before. I stood on my bed to tentatively touch it. After the scorching water in the Dorchester Center upper hall, I was hesitant to face another week of a blotchy cheek. I reached out a tentative finger and then pulled back right before I touched the flat surface, as if I had hit flame. I stood unevenly on the mattress, getting up my nerve, like I do before I kill spiders. I briefly considered whether I should use my left hand since I write with the right. Finally, I jabbed out a quick, left-handed finger, but the wall was dry. I pressed an ear to the plaster to listen for water, although the heat was off. The pipes were silent: no clanking, no dripping, and no rattling. I removed a framed print from the wall and stood across the room to discern a pattern. The mark was about four feet long and had a slight twist to its tube-like form. It was about six feet up the wall and seemed to have no obvious source. I was baffled. I don't think there are any radiator water pipes in that wall; it's an interior wall with no radiator on it.

It was very close to my bed; the wall bubbled up as I slept beside it.

Sometimes metaphors are confusing. Sometimes the mystery's too big for us. Certainly it's too big for me. What message could a rusted snake form mean? It looked a little like the history book's sketch of the sea monster chasing the Spanish ship.

Maybe my friend Tom with the Adirondack cabin was right; maybe everything doesn't mean something. Maybe I'm wrong to try to assign meaning to every moment. Maybe mankind is random. Maybe I just own an old house. Maybe the stain has been building up gradually, and I just noticed it for the first time. Every couple of months, I find plaster cracks in the living room stairway that I swear I never noticed before.

I blinked my watery eyes at the rusted vision in my sunny room.

I can't afford to paint my bedroom, I thought.

Throughout the day, I stopped working to compulsively check the stain, hoping against hope that it would disappear. From the windows, I noticed an unusually large number of birds in the maples. A flock of crows settled around my house. Whenever I walked outside, dozens of them flew out of trees and bushes, re-settling into neighbor's lawns. Later, as I wrote that night, the birds sang at midnight. Dogs barked under the avian chorus. Night became day. Why were they awake and singing? Were they talking about the stain on my bedroom wall? The animal and insect kingdoms communicate amongst their own species. Bees dance their excited reports of high pollen areas. Like humans, birds hand down stories in song, although most of the lyrics to their folk songs are "Robin here" or "Cardinal here."

Maybe all mythology is only saying, "Human here."

The neighborhood dogs yapped enthusiastically. Like the *101 Dalmatians* Twilight Bark, the dogs of Hamilton correspond at dusk, but lately, that relay of yowls seemed to continue far into night. Animals listen to their sixth sense; they recognize spirits. Eastern Shore Kent County superstition says that dogs howling at night is a portend of impending death. If you hear dogs barking at night, you should place your right shoe on its side with the upper part towards the dog to keep death away. I own way too many pairs of shoes for that spell.

It would take me hours to turn all the right ones.

> *Cambridge residents tell a story of an evil house on the outskirts of town, where a red-haired, eccentric old woman lived. Nothing grew in her yard, and dead trees ringed her property. She never bathed, and people claimed that she ate dog food and owned hundreds of cats. One breezy fall day, the delivery boy was dropping off her weekly dog food shipment, and she didn't answer his call. He thought she couldn't hear him over the wind, so he opened the back door. He found scores of dead dogs and cats, filling the rooms of her crumbling clapboard house. He almost couldn't continue for the stench. Every room had several dog or cat corpses in it, except the room where he found her twisted body. The coroner said that the old woman and all her pets died simultaneously and all of them died of heart failure.[51]*

Recently, Frederick County officials removed 184 dead cats and 100 live ones from a Mount Airy home. 284 is too many cats, still, I like the Cambridge dead cat story. Its ending gave me hope. All her pets helped the old woman cross. Isn't that the scariest part of the unknown transition of death – that we will all do it alone? The multitude of dead pets story was the second bit of ghost folklore that cast animals as transition guides. In Flowers' story of the two elderly residents who disappeared into the swamp, their mud tracks ended beside the body of a dead pet. Transition guides exist. A half hour prior to my great-grandmother and my grandmother's deaths, the air in the house sparked with electricity. It felt harder to breathe as if it was thickened with humidity. A half hour after they died, the additional energy had melted away. I like to think that someone came to help them cross over.

> *Once, on his way to work, a strange dog followed John Palin of Vienna for 500 yards and then disappeared. John knew every dog in Vienna, and he had never seen this dog. It was abnormally big, and it emitted a very earthy smell. When John arrived at work, he learned that his employer had died that morning.*[52]

Locals said the giant dog was a sign of the employer's death. Was the earthen dog vision causal or coincidental? Was it the employer's transition guide heading home? Vienna sits on the Nanticoke River in eastern Dorchester County, and at the peak of its population 420 people lived there. I suppose it's possible to recognize all its canines. In our endeavor to explain the unknown, we fall into the human trap of blaming coincidence. We blame misfortune on whatever precedes it. We are so linear. Our obsession with time makes us that way. Einstein developed the general theory of relativity in 1915, and we still think time is rock solid and lined up like a freight train.

"I don't have the time to come over," Karl said to me coldly on the phone. "I have another early day tomorrow."

A tear slid down my cheek. I didn't know why I was crying. Maybe I was leaking again. Maybe something in his voice sounded so far away. I felt the need for a transition guide, but I'm allergic to dogs and cats.

June 8th

THE WHAT

I've had the strange sensation in the past week that
my skin's absorbing more water than it usually does.
In the shower, I sense water pushing into my skin and
through my capillaries; I expect it to itch but it doesn't.
After all, I don't claw at my skin to scratch the sensation
of blood flowing. Last night, I turned off the spigot,
opened the curtain, and I was dry except for soggy hair.
I felt full and stretched; my pajamas fit tighter. I've been
much more thirsty than hungry lately; I've been drinking
gallons of water.

Did Big Liz feel this bloated, rising out of the
swamp? I thought.

A few months ago, my gynecologist told me that I
was on the brink of menopause and could face ten years
of hot flashes and night sweats, but last night was the
worst. I woke up, covered in sweat. I changed clothes.
I pulled the sheets and slept on beach towels. This
morning, when I dressed, my jeans hung on me, as if I

had lost five pounds to the night. It must be hormonal. We humans are eighty percent fluid; we're temporal bags of electrochemical water. Still, this odd leaking embarrassed me. Good thing Karl hasn't been around much to notice. I am pulled by the tides. And why not? As a woman, I'm tidal in other ways.

The office phone rang. I hung up the moist towel and answered. It was Maynard, Chief Winter Fox, head of the Eastern Shore Tribal Council. My hair was damp; my skin was pink and dry. He was soft spoken. He listened to the lining of my heart; he sent out tiny, tendril spies, searching. My aura was invaded. I intrinsically knew that the only way to allow him to talk freely was to completely open myself. I consciously let walls fall. I tried to just be. Chatting with him was like conversing with Yoda. I had met a grand master, and I was supposed to interview him. I felt at school.

He told me stories. I was so glad that he did.

"All Indians are the same people. The white man arrived in the New World and said, 'What do you call this river?'

'The Choptank,' said the Native.

'Oh,' said the white man, you must be a Choptank, then.'"

I giggled at the literal, European compulsion to categorize. He didn't say anything. I wondered why an Indian would mind being named after a river.

"The people who lived by the Choptank were easier

to buy out," he said, admitting to some "separatism" among Colonial Native Americans. He was referring to the Choptank tribe who in 1702 sold the last Indian reservation in Maryland to merchant John Kirk for "forty-two match coats."[53] A Cambridge librarian told me that match coats were beaver pelts, but I read that a match coat was a cloak-like garment made of fine European broadcloth and that the Choptanks traded forty-two of those cloaks for 3,000 acres of land. They had little choice. There were too few of them to be "well-organized enough to pose a military threat,"[54] and they had been trading with the English for eighty years by the turn of the 18th century. After the sale, they lived on a 16,000 acre tract east of town until William Vans Murray recorded the last one, a female ruler of the Locust Neck Indians, dying, in 1792. The smallpox and the booze got to their broken hearts. Native Americans believe that land cannot belong to a person. People belong to the land on which they were born. If a Navajo wanders too far from their native land, they lose their direction and purpose.

And if the Choptanks sold the last Maryland reservation, hadn't they been resisting sale up to that point? I wondered.

"It's hard to know history," I said about the match coats.

"We know ours. We Native Americans tell stories."

"That's a beautiful thing but no one in town's

talking," I said.

"Well, welcome to Cambridge."

"At least not talking ghost," I continued, speculating on the lack of ghost stories on High Street. "The white people of Cambridge wear blinders to the spirit world," I said. I didn't mention that I'm white; he must know that I am.

"There's a cost and a discipline attached to seeing the spirit world," Maynard responded.

I had heard that before from Korinne. She meditates on other reality possibilities, not opening up to ghosts per se but to all the energy of the universe. I don't think I'm strong enough for that daunting exercise. I've always suspected that revealing myself to the spirit world without a guide would be like spelunking in a dark cave without a flashlight. Imagine the monsters waiting in that primordial dark. There are enough monsters here in this reality.

"I'm reluctant to open up to you," Maynard said. "I don't know you. Are you married?"

I was surprised by that question; my romantic heart had nothing to do with these ghost stories. I didn't want to be defined by my marital status, but I had to share something personal for him to trust me. "I'm dating," I said tentatively.

"What does he do? This man you're dating."

Why does work identify us? Why does my boyfriend's work identify me?

"He researches."

"His work's very important to him," Maynard said.

"Yeah, I guess so," I said, wondering where this disconcerting line of conversation was heading. It took every ounce of my courage not to build up walls. I think we both were working towards that noble goal.

"How did you meet?"

"Through friends."

"What's his name?"

"Karl."

"Carl."

"Spelled the Czech way, K-a-r-l. With a K."

"One of my responsibilities as head of the Council is counseling Native American couples. I take it very seriously," the Chief said. For a terrible moment, I was afraid that he might try to counsel me. I had no intention to marry Karl. That's what I told myself. "Where's Karl from?" He asked, slipping into therapy mode.

"Well, uh, he's from Washington," I said, dreading counseling. "What about your lineage?" I asked, deflecting. I wondered why he was wearing European names like a match coat.

"I married into the LeCompte clan, one of the original thirteen families, although an influx of foreigners is shifting the tree some," Maynard said. He didn't answer my question. He called outsiders "foreigners."

I wondered how he fit into his own society. Eastern Shore Indian civilization was matriarchal and based on kinship. "Kinship determined whether one could become a chief or probably a priest. Kinship determined whom one could and could not marry."[55]

I didn't ask him about his mother. I told him that I was searching for a word to define state-specific xenophobic behavior, and he said, "For Native American Indians, xenophobia was a matter of survival." They had to stick together to maintain their culture, and they use folklore to continue those traditions. Until 1792, the four Indian tribes of Maryland's Eastern Shore dug into their reservations to maintain "semiautonomous social and political units outside English colonial society."[56] Beyond that, they kept telling and re-telling their stories, including ghost folklore. They had to.

Maynard fervently rattled off some failed American marriage statistics.

Where was this going? I wondered.

"A mixed marriage is a lot of work. More than a normal marriage," he explained. "It's hard work to be in this family."

All families are hard work, I thought. I felt oddly sure he heard me.

"But this family, oh, this family," he sighed, as if he had heard me. At a recent family gathering, there was an old lady that LeCompte claimed was going blind because of an ancient Native American curse. "Your people put the curse of blindness on this family, my family," she accused Maynard. "And it's not just the men who suffer."

This was not the first time he had heard the 350-year-old legend.

> *In 1659, Lord Baltimore granted Antoinne LeCompte, a transplanted French Huguenot, 700 acres on the Choptank near Cambridge in Castle Haven. The Native Americans, who didn't share the same European concept of land ownership, lived on his shores. Although the Natives displayed no violence and LeCompte's plantation manor was built like a fort, he was paranoid and convinced that the Indians would overcome his family at night. Historian Elias Jones chronicled that Mr. LeCompte fortified his new settlement for its protection with three-foot walls and heavy oak doors "and when the savages came menacingly near, he would disperse them by firing his large guns, and it is said, killed some of them."[57] According to several folklore versions, LeCompte invited the Indians to a sumptuous feast and served them plenty of rum. One by one, LeCompte's indentured servants lured the natives to the barn and murdered them. One brave Indian realized the plot and demanded to face LeCompte before he died.*

> *Confronted with his own demise, the Native American said to the gentleman farmer, "If you take my life this night, your children will not see the light of day."*

> *LeCompte murdered the brave or maybe*
> *his servant did the gruesome deed for him.*
> *From that day forward, every generation of*
> *LeComptes has suffered blindness.*[58]

"If you want to hear the curse first hand," Maynard said, "You should contact the side-shoots, not the direct line."

I assumed the direct line would fabricate, like the Cambridge residents, and like me.

I wondered how the white servants lured the Indians into the barn. More rum? Rifles?

Jones doesn't mention the blindness curse in his *New Revised History* book, but he does document that "Moses LeCompte . . . became blind when about twenty-two years of age. He was carried to Europe by his father for treatment but could not be relieved. Of his eleven children, nine of them lost their eyesight. Of the descendants of this branch of the family, forty-two became blind. In 1819, nineteen then living were blind."[59] Nineteen generations after the initial curse and eighty-one percent of LeComptes were still afflicted. The cautionary tale is a myth based on continuous history and describes a natural phenomenon, like thunder or lightening. It's a story justifying a mystery of nature. It's a metaphor of white men blinded by Native Americans that fits snugly into America's collective guilt of our ancestors' abysmal treatment of the original residents.

"Do you think that's the last time that happened? You think it won't happen again?" Maynard's raspy voice dropped into his chest as if he was telling a ghost story. His whispers scared and excited me.

I remembered the story of the 1840 revenge of the Shawnee Indians on William Henry Harrison. After the Battle of Tippecanoe, the Shawnee Chief cursed American presidents. He threw twenty stones into the fire and swore that each generation would lose a leader. Seven presidents in twenty-year increments died in office: Harrison, Lincoln, Garfield, McKinley, Harding, Roosevelt and Kennedy.

I didn't mention that curse. I didn't say that macular degeneration is genetic. I didn't tell Maynard that one Cambridge librarian told me that the LeCompte curse was sleepy eye, not blindness. I didn't tell him that I worked with several LeComptes in Baltimore in the 1990s, and they all wore thick glasses. I didn't mention that the curse could blind his own children.

Instead I only said, "Oh, I feel sure it can happen. Even to the side-shoots."

Some white, side-shoot LeCompte cousin was tracing the family genealogy and discounted Maynard because of his Indian blood. "He said to me that there's family you claim and there's those you don't," he said. "That's the people of Dorchester County."

Like me, he's a foreigner in his own state.

Like me, despite a separation from the other locals, he loves his marshy homeland.

Unlike Maynard, my ancestors were not marched off their land and herded onto reservations. My ancestors' bones were not carted off to the Smithsonian. My father's Norwegian people still sleep soundly in the Minnesota dirt.

Unlike me, Maynard's white side of the tree confidently claims the third original land grant from Lord Baltimore. He seemed pretty proud when he said that. Maynard's genetic tree is broad; its roots reached many soils.

When I mentioned Harriett Tubman, Maynard said, "She used Indian paths and hideouts. Blood's mixed between our families." He's the hereditary result of generations of blending civilizations. He's better for the mix genetically but culturally it must be tough.

Maynard grilled me on Native American culture; I knew some. I knew that the Indians believed that the deity created the world and split, leaving active spirits or *quicosock* that were manifested in nature and to whom the humans offered sacrifices of food and tobacco.

"Do you know about the six directions, the great hoop, the color of north, a break in the circle, the significance of the pipe, and the center of the world?" He rattled off a laundry list of religious rites.

"The six directions sound like physics," I muttered, taking notes and thinking of Einstein.

"There's mysticism in Native Indian religion; that's why I can be a Roman Catholic," Maynard said.

I tried to wrap my spinning head around that statement; I suppose that ritual is ritual, whether you believe in the six directions or purgatory. The six directions are the four cardinal points (north, south, east, and west) plus zenith (above) and nadir (below). I tried to place the six directions into the four dimensions of space and time. Up and down could equal north/south, side to side could equal east/west, and nadir and zenith were the earth below and the sky above. It made some weird sense.

"It's not cool to believe in the old ways," Maynard said sadly.

I understood; it's no longer cool to be a feminist.

"I speak for me. I carry the pipe," he continued.

I like people who speak in metaphor from time to time. People who speak in metaphor seem somehow on a higher plane of conversation and civilization.

Suddenly he asked, "Do you know what the Veil is?" The Veil, the Maya, the filmy, flimsy, fluctuating boundary between our perceived reality around us and what is really out there, the holey blanket between us and the Truth.

"Yes. Lord," I said. I'm not completely white and hopeless, I thought.

"Have you crossed the Veil? If you have and you are centered, then the beginning has no end."

I hesitated. "I think I might've crossed the Veil. Or stood on the edge of the Veil." How did he know? I wondered.

"You'd know if you had. You can only completely comprehend if you cross the Veil." There was silence. He was searching; I could feel him probe my heart.

I had stood at the edge of the Veil.

"My grandmother had diabetes and would get blood infections and slip into a coma unless she went to the hospital. She got tired of going to the hospital after years of it and told my mother that right as the last fever set on," I said, trying not to cry. "It was raining the day she died. She was sitting in her chair in her bedroom, and she looked like she'd fallen asleep. The neighborhood nurses wanted to resuscitate her, but my mother and I wouldn't let them. I knelt before her and held her hand and told her to go. Go, go, go. I thought I was screaming, but my mother heard nothing. I think I closed my eyes. I saw myself in a tunnel and I was pleading for her to go and I was letting go of her hand and there was light at the end of the tunnel. She melted into that light, became that light. It was a very beautiful place at the end of that tunnel, more beautiful than any garden on earth, beauty for which we have no words."

He listened. He breathed. He waited.

"Death is a transference of energy," I said.

The first rule of physics is that energy can neither be created nor destroyed. She did not end; whatever distillation of soul that was once my grandmother was not completely destroyed. Maybe I dissociated myself from a traumatic experience or maybe I'm in denial. Maybe I saw the Other Side of the Veil. The last lesson she taught me was her best: there is something after death.

"I don't know what's in the garden and I don't need to know, but I know there's something," I rambled.

"Why didn't you go to the Light?" Maynard asked.

"It was about letting her go. It was her crossing. It wasn't about me." Couldn't he see?

"Why didn't you go to the Light?" He knew the real answer. He could look inside me, behind the curtains and under the bed.

"I wasn't ready yet," I admitted, stuttering.

"Maybe those on the Other Side of the Veil didn't think you were ready either."

"She died eleven years ago," I said, back pedaling. "And it will take me weeks to . . .what could I possibly do to get ready?" I felt walls going up. I felt lost with no sense of purpose and direction. I was almost in tears.

"You should look up Bloody Henny and High Sheriff Scarborough," Maynard announced loudly. I had earned a clue. His clues dropped into the conversation, like small gifts parachuting down from the clouds, baffling boxes full of loose puzzle pieces. "Do you know who Sheriff Scarborough was?"

Something didn't sound right. "No," I said,

swallowing sadness.

Maynard said that Scarborough originated the phrase "the only good Indian is a dead Indian."

I couldn't believe he repeated that. "That's dreadful," I muttered. Then, it hit me. I realized that High Sheriff Scarborough is another name for the Sheriff of Nottingham, Robin Hood's nemesis. I felt a teensy bit played, so I shifted gears. Purpose returned.

"I saw two men," I said, "Older fishermen, they looked like, dumping fish, it looked like live fish, like part of their catch, into Cambridge Creek."

"You knew *quicosock*, a water spirit," he said.

Apparently, some local Native Americans still sacrifice to the water spirit in the river. After a fishing expedition, a gift of thanks to the river god is necessary for the bounty of future catches. An Indian myth of a water spirit that's half god and half demon and that lurks under the waves probably developed to keep Indian kids out of the deep end of the creek, and that myth of the Great Animal Master is still powerful enough to make grown men sacrifice part of their catch. Some say it's a giant snake, the Nessie sea serpent of the Chesapeake Bay. The Loch Ness monster evolved over the years into a collection of stories of mysterious wakes, humps and boat-less splashes in the Scottish Loch Ness. I thought of the first day I drove to Cambridge and saw that unexplained wake move sideways across the Choptank.

Could it be true? I wondered. The creek stretched

from the river all the way to the courthouse lawn.

"Could the water spirit have anything to do with the rumor of the Native American burial ground under the courthouse jail?"

"It's not a rumor," Maynard said, confirming Winslow's slip.

I waited, afraid to speak and break the spell.

"The burial was an unusual ceremony," he whispered, hoarse again. "Something about the way the burial was laid out was mystical and still is sacred."

Sacred is a very big, important word to a Native American chief.

Traditional Native Americans worship natural, sacred sanctuaries like rivers, mountains, lakes, caves, unusually shaped earthen mounds and rock formations for their strong earth power, but ceremonial and burial grounds can also be named sacred sites, regardless of their geographic significance. Ceremonial stories describe great revelations that transpired at the sacred sites. Mythic stories describe the sites as places that are "essential to the entrance to the next life."[60] Some sacred spots are what the Native Americans call a Thin Place, a place where time breaks through from alternate dimensions.

I wondered if time split into the six directions.

At sacred Hopi energy portals to the Other Side, compasses go awry and cameras photograph bands of orange and balls of light. Scientists can't classify the moving energy as radio, magnetic or electrical waves.

"I know there were ossuary burials, big open pits for the tribe," I said. The magnolia outside my office window was turning from blossom to leaf, and all its dying blossoms fell to the earth and stank. "I know that families were buried with their valuables and that the chief got his own . . ." I started.

"The thing they dug up under the jail *is a what, not a who,*" Maynard rasped, his voice scratching and tickling my ear.

Could I have heard him right? I wondered. He was using present tense.

"The What's not a Chief because the chiefs are buried in special log burial huts," Maynard continued cryptically.

Will someone build a special log hut for him someday?

I felt like a child, sitting at a lonely campfire with this man, listening in delicious terror to ghost stories of the dreaded What. Goose bumps raised on my arms. Everything stopped as if the planet had clicked into the right place. The magnolia stood still. The breeze held its breath.

The What resides in a sacred site, a place under the jail that is possibly an entrance to the Other Side. My mind was whirling, but Maynard kept talking. I tried to take notes.

"The pot-diggers moved them. And scant records

remain of those unlucky pot-diggers," he said. "Look around 1882."

1882. When the Church burned to the ground.

The pot-diggers, the white, influential people of Cambridge, were responsible for building the jail. They were the fools who disturbed The What. They opened Pandora's Box.

The Greek myth of Pandora tells the story of an Eve-like character, the first woman, who was given a forbidden casket by the gods. The gods ordered Pandora not to open the chest, but her curiosity compelled her to lift the lid, releasing plagues, sorrow and mischief to all mankind. Along with all the grief in the treasure chest was also Hope and "it remains to this day mankind's sole comfort in misfortune."[61]

I pushed away the image of Pandora running from the plague. I hoped I didn't get suspicious of The What story; Maynard could smell it. Maybe he was selling me a yarn to frighten the white man. What could The What be? My mind raced. The What could possibly be: an alien, a river or water spirit, a malformed child, a shaman, the Chesapeake Nessie sea serpent, a werewolf, the Yahoo monster, a mastodon, or a vampire. I'd uncovered the Area 51 of the Eastern Shore, and I found myself on the ground floor of a new myth. I vowed to tell my friends and family and to put the story in the tour; this is how legend perpetuates. Clearly, I was right in never seeking out Big Liz in the swamp. Clearly, the big story in Cambridge was on High Street. The biggest story was *inside* the town next to Spring Valley.

I was ticking through the outrageous monster What possibilities when Maynard rather randomly announced, "I can see the center of the world from where I live."

"Is it part of Nature?" I asked, scribbling madly.

"Ah, ha!" he said. I must've passed some test. He paused, and I could feel him breathe. He waited for my chest to follow his pattern, and we breathed together. His voice lowered into his stomach to relay my Quest. "Find the burial records, then call me back. Then maybe we'll meet and you can come to visit me. Bring the Czech." He said it twice, as if it was a spell.

Or did he say: bring the check? I'd rather pay him than bring Karl to him.

"Um, where are the records? The courthouse?"

"Have you talked to Tom?" Maynard suddenly asked, sounding like Father Martin.

"Flowers?" I had forgotten how a small town talks. I told the Chief the tomato gardening story Flowers gave me as his excuse for not discussing ghost stories.

"Don't think it's you," Maynard said. "That's what he's been telling everyone for the past two or three weeks. He hasn't been talking to anyone since his wife passed. He doesn't talk to any of his old friends. He's an unofficial mentor of mine, but he's not talking to any of us. It's not you."

I thought of how cats go away to be alone right

before they die. Suddenly I was worried about the old man.

Years ago, Flowers convinced Maynard to stay in Dorchester County and not leave directly after high school. "Stay and try to change the place from the inside, he told me," Maynard said wistfully.

"That's beautiful," I said softly. Maybe the old man was Maynard's teacher.

"He gets so tired so fast. It scares me," he said.

In his history book, Flowers wrote that the Choptank and the Nanticoke Indians eventually left Maryland by the mid-19th century and joined the Iroquois in Pennsylvania.[62]

According to Jones' history book, in 1672, Flowers' namesake served in the Nanticoke campaign, fighting the Indians. I wondered if Maynard knows that, but he must, because apparently, the Beginning has no End.

So, are we always in the Middle?

"I can set you up with an interview with her," Maynard said, announcing the end of our conversation. "With the old blind LeCompte lady."

You won't, I thought. I'll never meet you.

I thanked him for his time and hesitated. We both inhaled. I took a deep breath and told him the story of Ellen's joke about Baltimore writers landing in an unmarked grave. "I'm already folklore," I said.

"Oh, we wouldn't do that. We'd take you out to the marsh that knows is no bottom," he said in his deep

Quest Voice.

I tried to laugh past the image of drowning in gritty, stinky black mud. I tried to cloak my fear from him. I tried to hide my attraction, but his voice carved new lines down my throat. Maybe that's what happens when you let anybody into your soul that deep and far and fast. Maybe he found something in my heart I didn't know I had. I realized that my hair had dried during our conversation.

"Remember," he said, "there's a rhyme and reason for everything."

I slowly hung up the phone and researched some Native American terms. I found a definition of a shaman as a person who "due to illness, dreams, visions, or some inborn sensitivity or need directly experience the presence of spirits."[63] I realized that Maynard's a shaman, a medicine man, an intermediary to the spirit world. Shamans by definition travel back and forth from the land of dead where apparently there is good food and much singing and dancing. Medicine men can see great distances, foretell the future and look inside the body and mind. This man told me that he traveled to the Other Side of the Veil, and I believed him. This man looked inside my head, and I opened up my heart to him. This man could see the center of the world from where he stands, and it's a beautiful marsh. This man had warned me of its bottomless quagmire.

I paced my office for a few minutes and then

wandered to the backyard to think and to process all the belief-system-shifting thoughts that Maynard had planted in my heart. A shaman told me that the Other Side doesn't think I'm ready yet. A Water Spirit lives in the Choptank. The thing buried under the jail was A What, not A Who.

The daffodils and the hyacinth drooped over, tired, past their peak. I understood their malaise. I wanted to lie down too. It's tiring, getting ready for the Other Side. I sat beside the Cambridge tomatoes and concentrated on my breathing.

We've learned everything by watching nature. All our big concepts come from the clouds, hail, lightening, mountains, rivers, thunder and the sea. The growth of plants taught us time.

The mutant Dorchester tomato plants waved in a summer breeze; they were thriving beyond their watery transplant. Their stem height and leaf count had almost doubled. Tiny, hard, green tomatoes had replaced the yellow blossoms. They were much advanced beyond the pre-existing Baltimore batch that struggled by the fence. It wasn't a sun exposure issue; the fence gets full sun and the side yard shade.

Maybe the Big Tomatoes came from a Thin Place.

A cricket lay on its back in the garden, under a tomato leaf, wriggling wildly. It was missing one leg and struggling to right itself. I sighed and turned my glance away. As I looked to the crisp blue sky, I thought, how cold, but maybe that's how God feels sometimes about us. I reminded myself that it's bad luck to kill a cricket.

I missed my grandmother.

My father, the family's photographic scribe, took 35mm slides of all of our adventures and trips. Once, in 1967, we were all huddled in the living room, eating popcorn and watching old slides from earlier trips, trips before my younger brother was born.

My younger brother did not like watching slides of trips before he was born.

"Where was I?" He asked in an imperious three-year-old tone.

"You hadn't been born yet," explained my mother.

"Well, I had to be somewhere," he said indignantly, assuredly.

I think we all laughed, but I understood the logic of it.

"You were in heaven," my grandmother told him.

That appeased him for several years, but he still would purse his lips when she said it, as if he didn't quite believe it, even coming from her.

In my backyard, I pursed my lips and stared up at the limited sky and thought, she has to be here somewhere.

June 9th

FORK-LORE

"He said the only good Indian was a dead Indian."

"That's crap. Sheridan said that first, Philip Sheridan," Karl said. "Civil War Union general who burned through the Shenandoah Valley and then went west during Reconstruction to fight the Indian tribes out there. He coined that phrase. Everybody knows that. The Chief's cribbing. You gonna meet this guy?"

"He told me to find the burial records of the pot diggers and then call him. And then, and only then, could we possibly meet."

Karl sat in his boxers at my desk, losing in Spider Solitaire and waiting for some Gulag website to download. "Baby, you do not have a high speed connection. I don't care if they tell you that you have DSL."

"He wants you to come." I put on my robe.

"The old Indian does?"

"I don't think he's that old. He said, 'Bring the

Czech.'" I smiled. I was pretty sure how he'd react.

"Bring the Czech! That's what the Germans said in the 1940s!"

I walked back to the bedroom. Clearly, I was not getting into e-mail for a while. "He's not going to kill us," I called.

There was a pause. "Where's he live?"

The Chief lives happily in isolation in southern Dorchester County on Elliott Island with only sixty-seven other residents. He's protected on two sides by the Fishing Bay Wildlife Management land and by the Blackwater Wildlife Refuge. Elliot Island is eighteen miles through bottomless marsh to any town; those eighteen miles might as well be light years to the next viable star.

"Eighteen miles away from any town, and not, like, say, Chicago is a town! Or like New York is a town," Karl ranted on, leaving the office and thumping down the hall. He's a city person. He distrusts land without sidewalks. Karl was convinced we were headed for a watery swamp grave if we traveled to Elliott Island, but I could tell his intellectual curiosity was piqued. "What's the nearest town?"

"Vienna."

"Oh, not like the Vienna in Italy, I bet. How big is it?"

"About 400 people."

"Where's the nearest cop?"

"I have no idea."

"Bring the Czech! Them's good eats!" He riffed on, his arms flaying, pointing. He was having a fine time. He seemed oddly happy with the concept of the two of us dying in the swamp. The Czech people have a rather dark view of the world.

I watched him flailing and thought: we won't go.

When I told Korinne this story, she said that Karl was living evidence of "fork-lore."

We laughed.

When I finally got on my computer, I searched the Web for High Sheriff Scarborough. There were over 28,000 results.

June 14th

PASTA PAPER

All the hard work of spiritual processing made me vaguely agoraphobic. I wanted to curl up in a ball in bed. I couldn't focus. I felt vaguely fevered. I lost my appetite. I forced myself up on one elbow and called my touchstone Korinne again. I wondered if I could make sense.

"Talking to the Chief made me feel like the universe is forcing me to grow up," I said in a cranky tone. "Again."

Korinne said, "Oh, I hate it when that happens!" She obviously understood lessons in patience and Veil Preparation.

"Say those on the Other Side of the Veil think I should spend some time getting ready to cross, fine, all right, fine. I like being prepared. I was a Girl Scout," I blathered. "But, is that the only message I get? We don't think you're ready? How's that helpful? And how do I do it?"

Korinne laughed. "Like my message that space and time are irrelevant."

"Well, time is."

"Well, yeah."

"I accept that, and I accept that the Veil exists," I ranted on. "I get it. Isn't that the first step?"

"Hadn't you accepted that already?"

"Had I? Had I refused the idea before?"

"I don't think so."

"Well, what else could they mean?" I demanded. We didn't know; we're humans here.

I sat in the June yard and stared at a pair of ants traversing the Cambridge tomatoes. The one-legged cricket had vanished. I should've killed him and spared him the potential agony of being eaten alive by a bird. The sun dappled the azalea bush and threw bouncing shadows on a new batch of hosta plants that had popped out of the earth overnight. They weren't there the day before and then they were. Is the planet that simple? A spider ran and stopped and ran and stopped in the shade, and I remembered.

> *I was visiting a hotel in Dana Point, California. My room faced the Pacific, and to the north side, along the cliff, was a tent set up for a weekend wedding. The tent was empty the night I walked out onto the balcony to finish my drink. In my periphery, I caught a flickering light inside the tent, and as I watched the light*

*grew until it played upon the shadows of figures
dancing inside the billowing walls. The figures
were waltzing. The ladies' long skirts flew and
the gentlemen's coattails spun, as if the Others
allowed me a preview of an existence beyond
this one, an existence that I was not yet prepared
to join. The wind blew, the surf raged and I
cried. I clutched the cold, iron patio chair and
cried. I finished my wine and cried. The light
in the tent dimmed on the dancing silhouettes
and was gradually replaced by unsteady
moonlight.*

*I checked with the hotel's conference
manager the next morning and no one living was
in that tent the night before. No electricity had
been run to the tent yet. The lights I saw were
not physically possible.*

I had suppressed that wrenching, several-year-old
experience. Maybe I looked beyond The Veil that night
on Dana Point. It sounds crazy, but I saw what I saw.
My eyes were in conflict with my logical brain, and my
hopeful heart, the bloodied rebel, sided with my watery
orbs. One of Campbell's many definitions of myth is
"a manifestation in symbolic images, in metaphorical
images, of the energies of the organs of the body in
conflict with each other."[64] Didn't witnessing ghosts put
organs in actual, and not metaphoric, conflict? The old
man couldn't believe his eyes when he saw his dead wife
at the front door. The hunters couldn't believe their ears

when the Yahoo Monster wailed for weeks. I couldn't believe that I watched the dancing dead, and yet I did.

I wish I had remembered to tell Maynard the shadow tent story. The memory of the dance urged me to call him back. Maybe that's part of my lesson before I cross the Veil: don't repress. I've been bottling up my spiritual experiences, just like the city of Cambridge. The Other Side exists. I wish I could understand even parts of it better. I wish I could begin to interpret its baffling messages. What significance did the watermark and the black shapes convey? What language did the hot breath speak? What conjured them all of a sudden? I had never stacked all my ghost stories together into one quivering pile. Was that a spell unto itself? Did acknowledging them create them? Was it dangerous or healthy to face them? I thought of my friend Joe reacting to the Big Liz story: "No more. Stop. Okay. More."

Okay. Fine. More ghosts. Bring them on.

As the ghouls settled into my office, objects changed. On the diary entry on the 30th of May, I chronicled my nightly light exercise and how I surround the house with luminosity in defense of the breathing dream. I searched that section in the diary today, and it's vanished. It's gone missing from both the electronic and hard copies.

When I told Korinne about the missing section, she said, "Oh, they just want access," as if their robbery was the most common of events, like finding a snake on your bedroom wall.

"They who?"

"I just know 'they.'"

"Oh. Well, tough," I said. "I'm putting it back in."

"You should."

"I'm going to."

"Good."

I put the story back in, and the next day it slipped into the entry on June 3rd, emerging into another chapter. How did it travel there? I can't trust my own revisions, but I'm supposed to trust the universe. I re-typed the account into the May entry.

Images have been leaving narrative and surfacing into photograph. I showed the photo of the Choptank River to my friend Adrienne who Bill Pullman checked out. "Here's the wharf at the one end of High Street," I explained. "It's not a wharf. It's more of a sea wall or river wall."

"When was this?"

"May. Couple of weeks ago. The day I saw the graveyard."

She took it from me. "God, that's strange," she said. She held it up to her glasses.

"You mean that they call it a wharf when it really isn't one."

"No, don't you see it? Look in the clouds." She pointed to the cloudbank lowering over the Talbot shore.

"See?" We hunched over the photo together. At first I saw a choppy river before a rainstorm. Then another image jumped out: the unmistakable shape of serpent in the clouds.

"Oh my God," I breathed. Tingles ran up my spine.

"You can make out the body, the head, the tail," she said.

"Oh my God. It seems like it has wings."

"How strange," she said, drinking her soda.

Do I tell her, I thought. Will she think I'm nuts? "You don't know the half of it," I finally said, the words squeezing out reluctantly. I told her about the watermarks at the Center and on my bedroom wall. I told her about The What, the swamp monsters, the pirates, the dead doctor, the resurrecting wife, the headless horseman, the giant pine ghost and the legless sailor. I explained about the mutant tomatoes. "So many archetypes are adding up, and I feel so close to figuring it all out but it's just beyond the reach of my little, twelve percent, non-shaman brain. Everything's connected, but I can't figure out how."

"Maybe we're not supposed to know."

"Yet we've been granted the capacity to question!" I squeaked.

"That is a bother. You don't seem too freaked out."

"The blindness curse worries me a little," I rationalized, "but maybe the world's really like the mind on acid, and we block all the madness out because it's too horrifying. We can't function if we see the ghosts all the time. Maybe things moving around and snakes images cropping up are part of that freakiness. I'm just beginning to see the freakiness."

"You mean like the bedroom wall mark?"

"That and the little black shapes that zip along my periphery."

"You see . . ."

"Yes."

"You might need new glasses. Can I see the wall?" Adrienne asked. We walked upstairs to my bedroom. Outside, the moon struggled through the rainy afternoon. We stood silently, staring at the rusted watermark. I slowly tore the photograph. "That's probably a good idea," she said as I ripped the image. She didn't look at me; she stared at the stained wall.

"Maybe we should bury the pieces in the backyard," I said. "Or burn them. If I burn them, then they won't show up later, somewhere else, intact."

"Maybe it's all just a coincidence. Maybe it means nothing."

"I saw a boat wake on the Choptank when there was no boat in sight. I saw old men dumping perfectly decent fish over board," I said. "Those bumps were on my cheek for days." I could hear the fear in my voice.

"Yeah, that is a little weird, even for you." She tilted her head as she studied the wall. The moon slowly rose. "You don't want to come with us to the movies?" She finally asked.

"I'm sorry. I hafta write," I muttered, scuffing at the rug.

"You haven't left the house in a week."

"I work outta my house. I go to the store. For milk and stuff."

"Did you go today?"

"Not today."

"Well, you know we're right up the street," Adrienne said. "Weird," she said again and hugged me and left.

I finished writing the skeleton of the tour, stringing together ghost stories into a dark, moody necklace of memory. The horseman galloped wildly, the old man floated by in a coffin, the drunken doctor careened up the street and the shrouded woman climbed out of her grave.

Which phantom guide best suited all the phantom tales? I wondered.

I didn't want to use a dry Thomas Nevitt tour guide; he'd only be dimensional if he were drenched. I decided that if I could find the missing Willimina research or get copies from Olivia, I'd write the tour from her perspective of the warning ghost. Her regret would balance the fear drenched through the other legends. I wrote what I remembered of her story, but I couldn't remember all the dates. I couldn't recall how long she had been married to Goldsborough before she died.

I took a break to cook dinner, and as I reached into

a kitchen cabinet to pull out a box of pasta, out fluttered the first page of her missing research. It was wedged between the Indian pilaf and the pine nut couscous. I stood stock-still, thinking so hard I could almost smell it.

How did that get there? In this dimension, how was that humanly possible? The stories were shifting. I lost the light wall anecdote and gained Willimina's history. Was it because I had shifted folklore?

"I will not repress anymore ghost stories," I vowed to the empty house. A few hours later, I found the other Willimina page, nestled in between two towels in my linen cupboard. I leaned against the tile bathroom wall, sweating, concentrating. Karl never moved paper out of my office. I was the only person living in my house. Baffled and startled, I called Korinne.

"Even for you, that's a little strange," she said.

"That's what Adrienne said."

"Well, she's right."

"Apparently, I'm supposed to use Willimina's voice."

"Apparently," Korinne agreed. "Her gifts have returned."

Willimina Smith Goldsborough, who married against her heart, is buried in Christ Church graveyard. Willimina's father, William Smith, the founder of Washington College, disapproved of her match to Dr. Thomas Cradock, although he was the son of an Episcopal minister. Maybe William Smith knew that

Cradock's father was the first of his family in this country and Cradock couldn't recite back thirteen generations.

The couple planned to elope, but on the point of departure, Willimina hesitated and said, "If I go, I have my father's curse."

Despite Cradock's pleas, she refused to flee. She couldn't live outside her society. He was devastated and never married. In May 1783, Willimina married Charles Goldsborough, a member of one of the original thirteen families. Willimina died in childbirth at twenty-eight not long after. Charles Goldsborough married twice after Willamina's young death; men of that time period often outlived several wives.

All the members of mankind are walking time bombs, but 18th century women had particularly short life fuses. Childbirth is a killer. Would Cradock's unborn baby have killed Willimina as certainly as Goldsborough's did? Cradock must've wondered that sometime in his long life. I was a little surprised that Willimina didn't haunt Cradock. Maybe she did. Cradock wore her portrait ring to forty years of Episcopal vestry meetings.

I wrote the tour guide character of Willimina on an instructing quest, like the Chief had burdened me with Quest. Willimina would recommend that the tourists live life to the fullest and follow their hearts.

That night I dreamt that I was standing in a desert

and many birds were singing. I could almost understand what they were saying. I saw a shadow fall across the sand and turned. A tall, male Indian handed me a pair of sunglasses.

"These might help," he said. He had Maynard's voice; it boomed inside my head. I put on the sunglasses and could understand the birds. They were singing, "Human, here, human, here!" I looked up and the desert was replaced by High Street underwater. The river lapped up to the porch steps of all the big houses. Ducks swam down the street. Boats with pairs of men threw live flying fish over board. The flying fish had tiny wings on their backs. The wind blew. Water spun around me. Down the middle of the submerged street undulated the form of something long. Major Tom clung to a front porch; he was very wet. He tried to speak but water flowed out of his mouth in a cylindrical rope. In the braid of water were these black letters: Find the burial records.

I woke up thirsty. My body temperature in my chest burned my lungs. The pillow was hot; the sheets wet. I was glad that Karl was absent. I had showered before I went to bed, and as I fell asleep my pajamas felt restrictive and my skin tight. After I awoke, both my pants and top were loose from all the sweating. I am a balloon of water, I thought, and I rise and fall with the tides. The eventual failure of our bodies is part of the life cycle, part of the mythic cycle of birth, maturation, independence, failure and death, but it's hard to accept the eventual end of it. This body's the only sagging temple I know.

As I changed my pajamas in the dark of 3 A.M., I swore I could smell the sharp tang of tomato leaves leaking through the cracked plaster walls of my old house. I could hide in my bed until the end of time, but the stories would still find me.

June 22nd

THE DEVIL MADE ME DO IT

"Tell me a story," said my friend Lisa, the Bel Air dancer and joke teller. She cut up green peppers as we talked.

"I'll give you a legend," I replied, drinking my wine. I spun the yarn of the old woman in the coffin in the hurricane.

"Wow. The arms out, the hair spread," she said. "Creepy." She smiled, rocking on the balls of her feet, like dancers do. People like the creepiness of the ghost stories.

"An old dead lady in a floating casket, that's quite an archetype," I said. Archetypes are contagious and climb like viruses into our cells, burrowing in and changing our perspective, guiding us, like the old lady guiding her drowning husband.

"Hard to forget, archetypes," Lisa said, pouring us more wine. "Tell me another."

I told her Hannah's graver robber fable. I've been

storytelling lately, testing the truisms of these tales. Their universal truths work; most of my friends seemed happier to hear their eeriness. The myths are associative; they encourage people to grapple with their mortality. The stories grant humanity new hope of the beyond whether they are told or read. Still, the round echoes of spoken folklore differ from the flat, linear alphabet of the printed pages. I needed to hear the trapped words of the written tour in someone else's voice other than my own, so Lisa read it to me in her living room. The ghost stories sounded clear and real, but some of the language needed simplifying. Amazingly, the tour ran too long. After all that worry of lacking legend, I had to prune folklore.

"I have to cut it," I moaned.

"Don't cut the floating casket with the old guy and his dead wife with the hair," Lisa said. "That's quite an image, the floating coffin," she continued, handing me back my copy. "Makes me never want to live on the Eastern Shore."

"It floods there a lot. Every week or so. Their basements are tidal basins, " I said.

"Poor devils," she said, popping an olive in her mouth. "Keep the ghost stories and cut the history some," she suggested.

"History and folklore overlap daily. History gives the ghost stories perspective," I complained.

"Maryland, people are not going to take this ghost

tour for perspective," she said.

She was right, so back at the house, I trimmed the history sections. By deciding what to reveal of Dorchester County's past in relation to its phantom traditions, I was ostensibly creating history and no better than those history tellers, the conquering editors who wrote the schoolbook copy of this country. I paced my upstairs hall, thinking. I still needed facts to validate the Indian burial ground lore. Could history help me find my place in the world? History doesn't "help us to cope with the problematic human predicament"[65] as well as myth, but I needed research to buy a visit with the Chief. I checked online for Dorchester Courthouse history in The National Archives, the Maryland State Archives, the Library of Congress and the Maryland Historical Society. I found nothing. Researching live at any of those sources takes an entire day. There's nothing to eat for miles around The National Archives, and the parking at The Library of Congress is impossible. With a courthouse contact, I wouldn't need to camp at The National Archives.

I called Judy and nagged her about the promised courthouse phone number. Stalling, she suggested, "You should contact Thomasine, the volunteer at the Cambridge Library Maryland Room."

"She who denied Chesapeake Bay pirates," I whined.

"Maybe she can suggest where else you can go to

research," Judy suggested.

When I called her a second time, Thomasine announced that she was going on vacation for two weeks in Arkansas. "We're camping in the Ozarks," she said.

"Lovely," I said.

"We won't be digging up any pirate gold!" She announced solidly.

"Right." I couldn't think what else to say. The Ozarks do not border oceans; it's dubious that its rolling mountains would harbor sea pirates.

"Go to the Maryland Room of the Enoch Pratt," Thomasine suggested. "If you can't find the Indian burial records at the Enoch Pratt, then call me when I get back."

I never called her back. Hers is not the path to knowledge. Why listen to more white Cambridge denial?

In the Maryland Room of Baltimore's Enoch Pratt Free Library, I discovered no information on the construction of the Dorchester County jail. All the maps there were no older than 1935. I needed burial records. When I asked the Pratt librarian about the courthouse records, she looked aggravated and asked me, "Have you gone to the courthouse?"

I sighed. Like Lincoln's dreams of his viewing, I knew I shouldn't go to the courthouse, but I didn't know why. Maybe because the people of Cambridge are closed clams. Maybe because I suspected that all the

Indian burial documentation had been burned, either by accident or with intent to cloak. Maybe I didn't want to go because a corner of the courthouse building was blown off in 1970. Maybe I feared the bottomless marsh.

"You could go to the Maryland State Archives," the Pratt librarian said, but the Maryland State Archives web site had no information on any Dorchester County courthouse building records.

The Enoch Pratt yielded no courthouse records, but I found several great folklore sources there, mostly compilations by George Carey and Vernon O. Griffin. Dorchester County ghost stories put the rest of the state to shame. The swinging chandeliers of Elkridge are barely creepy next to terrifying entanglements with the Devil in swampy Dorchester County. I deleted all but one of the tour's non-Dorchester stories when I dug up the local demon and witch tales. Thanks to Carey and Griffin, I had landed safely out of the fabrication woods. A drop of water fell from my eye to the page as I read the tall tale of Molly Horn. The guilt ran clearly down my cheek, like the stream tearing out of my itchy eyes.

Carey tells the farming story of Molly Horn, a fine example of a Dorchester myth, a parable of unknown source that explains the creation of an island from the mud shook out of the Devil's shirttails.

> *Molly was a shrewd woman, and she made a farming pact with the Devil. First she told the Devil that she would harvest everything that grew in the ground and the Devil would reap everything raised above the ground. Molly planted potatoes, so the ground yielded her a crop and the Devil only got potato greens. Next Molly promised the Devil everything raised in the ground and she would keep everything that grew on top. She planted beans, so the Devil only got roots. Infuriated, the Devil cornered Molly Horn on the Northwest River in southern Dorchester County. Molly hit the Devil, and he skidded across the marsh to the edge of the water. The Devil stood up and shook the mud from his shirttail and made Devil's Island and dove into the water and made Devil's Hole.*[66]

Like Ebenezer Scrooge and the people in Greek mythology, Molly, the fallible woman, calmly and rationally negotiated with the cranky supernatural. Myths cannot just tell tales of gods and monsters; human characters must struggle with the monsters so we can relate. Maybe Molly's lesson is that we can all stand up to the Devil, but why did she make a pact with him at all? Everyone knows you can't win at that.

What can one say about the Devil? He likes a red light. He plays cards. He's followed by a drumbeat that matches and wins your heart. He's sneaky and deceptive. He can convince you to lie and slide down that slippery slope to murder, greed, jealousy and fabrication. He wilts flowers. He's the opposite of light. And as the opposite of light, space and time behave differently with him.

Maybe the Cambridge townies are in denial, but the realistic marsh residents of Dorchester County recognize the Devil lurking inside themselves.

> *At midnight, Lloyd McCready was walking home in Ward's Crossing, and he saw a fire illuminating a drainage ditch by the side of the road. Drainage ditches are wet and usually not ablaze. He had a .22 pistol but not the strength to pull its trigger. He walked over to look in the ditch but stopped when he heard chains rattling. Recognizing the sounds of the Devil, he tore off into the night, all the way home to his mother. His mother said of the fire in the ditch: "That's probably some of your meanness coming out of you."[67]*

What's that about? Did Lloyd have a nasty streak? Certainly, I have unkind moments. We all hide our shared capacity for meanness, but the mother's question seems like a big, jarring story jump. Maybe Lloyd had gambling issues. Maybe our collective cruelty creates the Devil. He is in all of us; we have crafted him.

> *The next day, Lloyd checked the ditch for signs of the fire, but all he found was a deck of cards torn into a thousand pieces, for they say that the Devil is in cards.*

There's evil in the world. There's good energy and bad energy. There's karma; the Hindu and Buddhist construct that our lives are dictated by our behavior in this and previous lives. How do we remember the lessons from previous lives? Is that why we leave stories behind? As clues for ourselves the next time around? Was I reading clues that I had once left behind but had lost the knowledge to recognize them?

Shuddering in the cool of the Pratt library, I read on. Another didactic, ethical tale of the Devil sounds like Biblical parable, and clearly superstition grows out of this ghost story.

> *A man who lived down in the neck of southern Dorchester County craved gold. One night, he swore he would give his soul if he could have some. That night, a strange man appeared to him and told him to hang his boots upright against the wall and he would receive all the gold he wanted. So that night, the man hung up his boots, and the next morning they were filled with gold. Whenever he needed money, he hung up his boots, until he became so greedy that one night he cut the toes out of the boots so the gold would run out and pile on the floor. When the man awoke the next day, his floor was covered with gold, but that night he disappeared. People said that the Devil had come for him.*

> *Across the county, it is said that you can't hang your boots on the wall or something will throw them down in the night.*[68]

Why hang boots on walls at all? Was the damned man's eternal task to throw boots off walls? I wondered. If so, wouldn't he then be working for good and God? Wouldn't the Devil want the temptation of the boots up and full of gold? Did the man finally stand up to the Devil in Hell? As poor as I was, I was, frankly, too scared to hang up my Doc Martins overnight. I have enough devils of my own and greed is surely one of them. I'm human and in debt. I wasn't too sure that I could reject boots full of gold. I don't know about cards but the Devil is certainly in money. Everything comes down to money.

I suspect that Cambridge's ghost denial stems from its class and economic gaps. Most of the Dorchester phantom folklore bubbles out of the lower classes of the county, from the people who live closer to the land and weather, respect their elders and their stories, inter-breed enough to produce gullibility and live in enough isolation to restrict them from other forms of entertainment. I wondered if money blocks out the senses, if the wealthy pad a thick wall of fiscal safety around their petal-soft ears or if the sound of all that gold thudding to the floor is deafening. Maybe temporality is more real to those who daily struggle for bread. That daily fight requires some sort of definition, some reason for the great battle. We blame our bad luck on others, like those on the outskirts of society.

Since the Inquisition, communities have blamed

their ill luck on the older, single or widowed, healer midwives. The Dorchester County women accused of witchcraft hovered on the outskirts of society, scapegoat commoners not associated with great wealth.

> *A farmer noticed that his chicken stock was diminishing overnight and with every disappearance he spied a black cat. One windy night, he caught the cat and threw it into a bonfire. The following day, one of his sharecropper's wives was indisposed, taken to bed because all of her skin had mysteriously burned off her body. After the sharecropper's wife was scorched, no more chickens were stolen from the farmer's hen house.* [69]

I changed "sharecropper" to "neighbor" in the tour version; that former term might be in bad taste, outdated and Southern in a bad way.

After all, words are powerful, as powerful as spells, as powerful as beliefs. Maybe that was why Delia wanted to edit the High Street stories.

> *Another conjuring tale developed in Hopewell, Virginia of a woman who was suspected of being a witch and who disliked a waterman. In the story, the woman sketched the man and left her drawing by the water's edge. While the man was at work out at sea, the tide came in. As the tide washed over the woman's drawing, the waterman fell overboard and died.*

The most amazing part of this myth is the fact that most Eastern Shore watermen can't swim.

Folklore is as didactic as myth and suggests many defenses against witchcraft, including remedies such as a pan of cold water under the bed or sprinkling salt around the house. Some hang a broom over the door to deflect witches, because by the time the witch counts all the straws in the broom, it will be morning. Some suggest hanging a flour sifter over the door, for a witch is hesitant to enter because she'll have to exit through every hole in the sifter as she departs. But, brooms or flour sifters or not, it's not possible to always protect yourself, even if you nightly surround yourself with light. If a person wakes up bone tired, the Dorchester Islanders say, "The witches have been riding them."

> *A Cambridge woman told folklorist George Carey that her mother was convinced as a small girl that there was a witch in their community who didn't like her. The girl said that the witch had been riding her at night and not letting her sleep and that's why the child was waking up with sore muscles. The girl's mother told her to put a fork under the witch's chair at the next opportunity. When the suspected witch next came over to visit, the girl placed a fork under the rocking chair where she sat and trapped the witch. The witch couldn't move until she swore that she would no longer ride the child at night. She promised and the child removed the fork.*

That night the girl slept through the night for the first time in months.[70]

Why the fork? Did it represent a pitchfork? A trident? Choices that we daily make? A fork in the road? Was I being too literal? My arms ache sometimes when I wake up. Was the morning bed breath the leftover vestiges of witch? Maybe I needed a conjurer to provide a recipe to keep the evil at bay.

Dorchester County superstitions attempt explanations of the inexplicable and often those explanations evolve into folklore. We need stories to justify the irrational vicissitudes of life. The marginalized, independent women who rejected the strict, social rules of the "subservient female" were faulted with bad crops and dying cattle because they refused to marry or live with their families.

We fear difference.

Margot Adler, in her book on American paganism, theorizes that the "unity, integration, and homogenization in the Western world derive(s) from our long-standing ideology of monotheism."[71] Social, political and economic homogenization directly extends from monotheist religion. We are encouraged in American society to dress, talk, and act alike. In the monotheists' world, the eccentric becomes evil and the diverse becomes wrong. We are not encouraged to be different in America. There's a dangerous freedom attached to individualism and a pretty steep cost for that freedom. You act differently and you get labeled as crazy or witch.

Adler details the much debated history of Wiccan ritual: some believe that several core families have maintained pagan traditions; some believe the Catholic Church fabricated witch stories to justify the murder of Protestants during the Inquisition; and some believe that the Inquisition encouraged an existing obsessive delusion that began with folklorist Jacob Grimm and played upon man's darkest fears. We have all lost sleep as children, convinced that the wicked witch in the backyard would hunt us down and roast us alive, and folklorist Grimm fanned that fire.

The entire witch tradition could feasibly be based on Grimm fairy tales.

"We all know the story of Hansel and Gretl because parents in the Middle Ages didn't want their kids to go into the woods," said my friend Joan with her arms spread in an obvious shrug.

That's the far-reaching, shape-changing, gut-wrenching, sleep-depriving, crowd-lynching power of folklore.

According to Adler, the Latin roots of the word *witchcraft* lie in two words, one for *wit* and one for "the craft of shaping, bending and changing reality."[72] Witches are labeled as *shape-changers*.[73] Stories of shape shifting are as old as the myth of Petronius, the Roman soldier who turned into a wolf during a full moon. Adler says that the word *witch* itself is much more powerful than a single noun; it's a collection of ancient images that merge into powerful archetype. It's a wolf braying at a full moon and a circle of dancers in the forest.

> *My high school Advanced Biology class camped out in a Chesapeake Bay Foundation house in Bishops Head in southern Dorchester County. The town of Bishops Head was named for a decapitated bishop, or so the legend goes. My class was in Dorchester County to hike through Blackwater Wildlife Refuge that spans 10,000 acres of fresh water marsh between an estuary and a river. The refuge's really a big filtering system. As a refuge for thousands of birds, its sky is whipped by the whoosh of wings. Blackwater smells like the soup stock of the world, percolating and simmering, sinking into your skin.*

> *According to the Maryland Ghost & Spirit Association's web site, a green-eyed mule ghost haunts Blackwater. I never saw a mule, much less a green-eyed one, when I visited the refuge in high school. By day, we students mucked through the bog, sinking hip-deep in dreaded marsh holes. Walking through swamp is tricky. You have to jump between islands of solid ground, and if you miss, you quickly descend into thick, gritty sludge like black oatmeal. Once both of my legs sunk in, and I clawed madly for clumps of anchored grass. Two of*

my comrades pulled me out of the muck, but
I had not forgotten the cold panic of slipping
under the gluey mire. I recognized that old panic
in Maynard's threat about the marsh that knows
no bottom.

After supper at Bishops Head, several of
my friends and I went for a hike. After a good
half hour walk in the woods ringing the swamp,
we stumbled upon a clearing with hundreds of
footprints in a circle in the dark mud, around
the remains of a fire. The footprints were still
damp and were all heading in the same counter-
clockwise direction, counter-clockwise, like the
spinning of a hurricane. We stopped, silent. I
felt watched by the trees. In a flash, I visualized
hundreds of barefoot humans dancing around a
campfire and heard the echoes of their chanting.
The smell of our sweat changed to something
sharper and tangier. We were scared because we
had been taught the Grimm archetypes. We had
been taught that a gathering like that was wrong.

As recently as the 1980s, pagans danced in southern
Dorchester County.

Early stories of witches created an archetype
and that archetype, in turn, perpetrates other stories.
Broadsides sold on 17th century European streets were
pamphlets depicting gory tales of werewolves and
witches. Victorian England penny dreadfuls spread the
myth of warlocks and Jack the Ripper. The publishing
industry has a booming horror genre. Arthur Miller's

frightening play, *The Crucible,* about the Salem witch trials, is still being produced. The Internet circulates videos of seemingly normal rooms that suddenly are inhabited by a drooling phantom. Archetypes work, particularly on subjects we don't understand, and that's just about everything. I stared at the gray sky outside the Enoch Pratt library windows.

I know nothing, really, I thought.

I opened up George Carey's folklore books and found Bloody Henny. Maynard's tip panned out: she was an African American accused of witchcraft and executed off an ox cart at Spring Valley between the courthouse and the church.

> *People claimed that a local man hired Henny Furr in retaliation over a dispute about a cow. Henny brewed up a potion and put it into the victim's coffee, but his granddaughter drank it by accident and vomited and defecated snakes for three days. No doctor could heal her. Her cure was as curious as the curse. Folklorist Carey wrote that the girl's father went to "another local conjure in town,"[74] like it was once as common as running to the pharmacy. The other conjurer burnt a pile of nails with whiskey poured over it and told the father to bury the whiskey bottle in Henny's front yard. The father did so, and his daughter stopped passing reptiles.*

Usually snakes represent knowledge, sex, or energy

in myth. I thought of the water stain snake and the photograph cloud snake and shivered. And how could anyone live through three days of barfing snakes?

There are good witches and bad witches, and Cambridge once had a decent conjurer that could counter the wicked spells. I could use one now.

Modern pagans live by the vow of "do what you will/harm none." It's a splendid credo, but certainly not Bloody Henny's style.

> *When a neighbor's dog trespassed on Henny's property, people said that Henny turned the dog on its master, and it mauled him to death.*[75]

> *A farmer complained that something was ravaging his cabbage patch nightly until one foggy night he finally shot what he described as a "strange animal"*[76]*as it ran between the cabbages. The following day, everyone whispered in town that Henny, then nicknamed Bloody Henny, was mysteriously peppered with buckshot.*

Like the sharecropper story, the shape changer can transform from woman to cat, but, weirdly, injuries sustained in animal form are carried through the transition back to human. That injury carryover seemed oddly literal to me. Still, no matter the questions and inconsistencies, the power of those stories were enough

to hang poor Henny. Women were hung. In Europe,
60,000 to 300,000 accused witches were burnt alive
during The Burning Times, the two hundred years that
began in the late 1600s. In 1692, 200 Americans were
either hung or pressed to death in Salem, Massachusetts
for witchcraft.

Everything is connected, like the Native Americans
believe.

Witches could be based in folklore. Shamans have
been accused of witchcraft. Chief Winter Fox is a
shaman. The What is buried in a sacred site under
the jail. A sacred site can be a door to the Other Side.
Bloody Henny was executed for witchcraft on a sacred
site. Snake images were cropping up. Fishermen were
throwing back their catch. I'm turning to water as I
write a ghost tour about this watery, sacred site, and
I keep finding more stories, more reasons that High
Street's a kind of vortex, or maybe once was the center
of some sort of big energy. The whole town's in denial.

I can't stop thinking about ghosts, witches and sea
serpents. Besides the nightly light visualization, I placed
brooms at each entrance to my house and considered
hanging my flour sifter over the kitchen door. I stopped
at scattering salt and putting a pan of cold water under
my futon. The salt suggestion is too messy, and I'm
confounded how to keep the water cold under my bed
in summer. The metaphor of cold water under a bed

to block evil is rather mixed. In archetype, water is symbolic of emotion, dissolving, yielding release, and cleansing.[77] Would the cold water trap the evil?

Even if I attempted the cold pan of water, no one would notice it under my bed. I hadn't seen Karl in a week. He was fixated on researching some Czech guy who was tried as a spy in the United States in the 1950s. He talked about this guy constantly. He spent his days at The National Archives.

I'm obsessed with ghosts and he with spies. Our conversations made no sense.

We fought this week. I was imitating Olivia's Southern inflection, and he called me a racist because I emulate other people's accents.

"I'm maintaining the country's quilt of oral history, that's all," I said, pulling the car over. I couldn't have that discussion and drive the car.

"I don't know why you're so upset," he said.

"You called me a racist!" I screeched. "Certainly I compare myself to the world by comparison to the group of white Yankees who raised me, but I don't judge others on the basis of skin color or place of birth!"

"What's the difference?"

Maybe we are attracted to the people who we are really allergic to, like food allergies, like the attraction to the mold in blue cheese.

Did Willimina have this problem? Did Cradock love

his job more than her?

"Do you want to get out?" I asked. We were only about ten blocks from his apartment. I wished we were further.

He looked at me long and hard. I cried. "No," he finally said.

A sea inside me shifted and sloshed. I was cold and nauseated. Did Big Liz feel this way when her master turned on her?

"I'll call you this week," he said as he left.

I sat in the car, staring at a tree. I sneezed. Was he right? I wondered. Was imitating accents equal to racism or at the very least xenophobia? Maybe the Chief was right about us. Maybe we needed his counseling.

Bring the Czech indeed.

June 25th

I HEARD NO WATER

Two friends of mine have connections to people who investigate Native America burial grounds, but neither helped me with the courthouse mystery.

My playwright friend Ronda wrote a play about Native American Indian burial removals paralleling the mass graves of Serbians, but all she wants to discuss is Yugoslavia. "They're still uncovering mass graves!" She lamented, shaking her dark curls, wanting to tell the stories of her people. I met her in a playwriting class, and she's a librarian by trade.

"It's just wrong. All those bones carted off," I said. "All that unfinished burial creates bad energy."

My costume designer friend Denise knows a guy, Kevin, who's an Indian cartographer.

"You'll probably have to go to The National Archives to find the courthouse jail records," Kevin wrote in email. "And I could probably go with you."

After several emails, he rescinded his offer to accompany me.

"I only make the maps," he wrote.

I told Denise this exchange. She's a native Baltimorean and smells of patchouli. She grows her own herbs and belongs to band called The Dirty Mothers. She sees ghosts. "What changed his mind?" She asked.

"I don't know. And The National Archive database has no records on the Dorchester County Courthouse."

"Strange," she said. She and I know strange. She and I both had weird experiences in a bathroom in St. Mary's Outreach Center in Baltimore.

> *We were both alone in the bathroom at different times. The bathroom has two stalls, and we were seated. Denise saw a darkness, a black shape, go by outside the stalls, at foot level. I heard the unmistakable sound of someone opening the adjacent stall door, closing the door, removing clothing and sitting down. When I heard it, I looked under the wall and saw nothing. No feet. Did I really hear what I heard?*
>
> *I remembered Denise's darkness story and called out, "Um, I'm being really human here!" I heard the sound again of the stall door opening and felt the stall walls vibrate. Goose bumps rose on my skin, and I hurried to leave.*

Denise's fiancé Anthony experienced my version of

the peeing story in the adjacent men's room. Our friend Raine heard the other stall door open and close when she was clearly alone. Raine has a theory that ghosts are ripples in the energy of the universe. She's another red-haired Baltimorean. She's meticulously clean, completely beautiful and outrageously funny.

"Why would ghosts haunt us on the john when we are our most human and vulnerable?" I asked her. "It seems unfair. Maybe they miss being human."

"I find that hard to believe," she said. "Who would miss that?"

"I heard no water."

"Me either." She said, her sapphire eyes big.

"I wiped and pulled up my pants and booked."

"Me too."

Were more ghost stories happening to me as I wrote this story or am I just more aware of that which has been been pulsating around me always?

I told the senior ladies who ran the Nearly New Shop in the Outreach Center the ghost in the stall bathroom tale. Bunny, the Roland Park lady who runs the shop, went slack-jawed. "You mean that bathroom?" She pointed up the short flight of stairs.

"Yeah."

"Well," she said, adjusting her plaid skirt and her penguin sweater. "You don't want to hear that. Why would I want to know that?"

"I thought you'd want to know," I stammered. "Less

surprise to you when it happens."

"It hasn't happened to me yet," she pointed out.

Did Cambridge not want to know? The Cambridge population seemed split between those who did and those who did not.

Judy called me, pressuring me for an end of June deadline before she went to Charleston for vacation.

"If you really want the tour, you can have what I've written so far, but I think you're missing the best story in what is under the courthouse jail," I said.

"Oh, we shouldn't worry about that. We need to be up and running by mid July and the guides have to memorize it. There's a lot of it, you know," Judy said. "Oh, I have a courthouse name for you, and I also have another name for you for more ghost stories. It's a local couple named Travis who go to Christ Church." She gave me the contact information and left for South Carolina.

Dorothy was the contact Judy finally gave me as the courthouse connection, but her phone number was wrong. I called the main courthouse number and left two messages for Dorothy in the Register of Wills, and she didn't answer back.

I finally blind-called the Smithsonian. I wanted to go to them with more burial information, but I couldn't wait any longer. I left messages with an archaeologist and a curator named David Hunt.

As I dialed the Travis phone number, I wondered
why I was even calling. The only change I wanted to
make to the tour was further research embellishment of
the Indian burial ground story. Apparently, I was hooked
on ghost story. I wanted to hear more.

A woman answered the phone. I introduced
myself, and she said abruptly, "You'll want to talk to my
husband." She dropped the receiver with a clank. She
had experienced a ghost first hand and didn't want to
discuss it. Her husband explained.

> *They had just moved into a Victorian house
> in Cambridge, and the wife walked upstairs
> alone. In the upstairs hall, she encountered an
> older man in a plaid shirt. She thought he was
> a live intruder and didn't know how he got into
> the house. She screamed, "What are you doing
> here?" As she backed away, her husband called
> her name. She turned and when she turned back
> the man had vanished into thin air.*

> *Later, she saw a photograph of the dead
> previous owner of the house, and it was the man
> in the plaid shirt.*

> *"She said he looked so real," Travis said.*

Just like what Casey and the other Indiana techies
said about the Victorian woman. Just like the bed
breathing sounds and the sounds of the stall in the
church bathroom and the man in the hunting lodge

living room. They seemed so real.

Travis told me more of his boatload of ghost stories from his hunting lodge on Aisquith Island, on the southern belly of the county. It was refreshing to find a Dorchester County source that seemed proud of the rich heritage of the phantom myth.

> *A hunter staying alone in the lodge heard doors and footsteps and followed the sounds to the bedroom. He found nothing. When he turned, he saw the pullout sofa bed in the living room bouncing violently as if something was jumping up and down on it. The faucet in the bathroom turned on by itself. The hunter grabbed his hunting gear and bolted.*

> *"And he was a big man, 6'3"!" Said Mr. Travis, laughing. "And he had a couple of really big guns with him!"*

> *A group of hunters were standing outside the lodge and all saw a green glow exit the house. The green glow passed through the door, hovered over the porch and disappeared.*

> *"We had been drinking," said Travis, "but not that much!"*

The Travis lodge was built over an old Indian burial ground like the *Poltergeist* house. That kind of energy does not go away; it sticks to the trees and the grass. It paints the soil red. This universe is a closed system;

earth's atmosphere makes it even closer. Energy does not disappear; it just changes.

American demolition experts, the Loizeux family, acknowledge that the energy used to create a building is stored in it. When imploding a building, they harness that energy, turning it from stored to kinetic energy, to help bring the structure down.

How are haunted cabins different? How are haunted hotels different? How are haunted jails different? How are humans different?

Mr. Travis told me his tales and paused, dying to be immortalized in song and story.

"What have you seen?" he asked me.

"Some strange things," I said, slowly, still taking notes.

"Like what?"

I couldn't think of a single one. Maybe I had replaced them with Dorchester ghost tales. I had repressed them all again, despite my Other Side warning. The ghost moments have been slowly surfacing into this narrative, bobbing up like a river letting go of a corpse. Like the people of Cambridge and Easton, I had forgotten them all: the cable ghost in *Camelot*, the alien lights over the lake, the Dana Point shadow dance of the dead and the Colonial time warp. Maybe I couldn't remember them because I writing down Travis' stories. Maybe I didn't want to sound as eccentric as he did. Maybe I feared difference too.

"There was a time," I said, "when my stereo turned on in the middle of night by itself, tuned to a station I don't use." Maybe I wanted to keep a journalistic distance from my sources.

"Sounds like your house is haunted too," he said.

Most houses over fifty years old are haunted by their history.

Take the city of Cambridge, for example.

Judy had been in South Carolina for one day, and Delia called me, asking for the tour script. I told her the same thing I told Judy. The best story is under the marina.

"Under the jail's a good story, but we don't need it in the tour. It's scary enough as it is."

I caved and agreed to send the script; I was commissioned, after all, and on a deadline. "But I wish I knew the jail story. It's driving me nuts not to know," I complained.

"I might have a friend on the city council who might know but I don't know what help they'll be. It's hard to get people to talk. Maybe it's this town. It gets to people," she said in a manic rush. "The people who have been here over ten years; it's like a poison."

I didn't remind her that she was in her ninth year and Ellen the librarian was in her eighteenth.

"I have to go, my other line's ringing," she said.

I thanked her for the project and hung up.

I e-mailed her a tour version sans full What tale. As I pushed the send button, I thought, there it goes. It's over now. No more worrying about research. No more trying to find time to go to archives. No more repressive neurotic dynamic of a small town. No more thinking about ghosts.

June 27th

CANCELLED CZECH

I wrote down The Argument to see if it made sense. I used this diary as a tool to evaluate the reality of an end of a relationship.

Karl said I was too emotional. He said I had too many romantic expectations. He said I'm distracting him from the work that defines him. He doesn't want to find the time to work on our relationship. He doesn't think he has it in him.

He said, "I guess it has come to a close."

He said, "Clean breaks are better."

Did Willimina make the break between her and Cradock? I wondered as I listened on the porch to Karl's voice. But she wanted to be protected, she must've or she wouldn't have left the protection of her father for the protection of a rich Goldsborough.

"What did you say?" I asked Karl.

"Maybe I can't make people happy," he said. He said working on our relationship would be a waste of time.

He said he'd call me in a couple days.

I watched him walk away down my street.

Maybe he was right. I should break off all attachments and focus on the mystery of ghosts.

Dredging up all these ghost stories has been a muddy business. It's got me thinking big picture thoughts, and big picture thoughts have a tendency to make linear, logical thinking harder. It's a brain use issue; I only have so much brain to use.

In this past week, my boyfriend and I broke up, my cell phone died, my air conditioner died, my office was robbed and I lost a diamond earring. How long before I begin blaming this bumpy stretch of life on these ghost stories?

In Flowers' folklore book, a woman disturbed a Native American Indian burial ground by removing a talisman. No matter where she put the medallion in her house, it ends up every morning under her pillow.[78] How similar is that folklore to my story of Willimina's missing pages? If the talisman story happens when you move a little charm, imagine what can transpire when you move the entire sacred bone yard to the Smithsonian. Imagine what happened in the 1880s, when the white people of Cambridge disturbed a sacred Native American site.

My friend John told me this ghost story. John is a tall and lanky dog-walker, landscape designer and actor. He was born in Baltimore, and his voice is deep in his chest.

We sat on the porch the night that Karl left me, drinking
with Raine.

> *So I was talking about Kurt's ghost story,
> in Chris' basement, about how the phone
> company had dug up human leg bones and left
> them in a plastic bucket on the steps and how
> all sorts of stuff starting happening after that.
> So, Chris and I are in his basement, right, and
> Chris had all these wall hangings, remember?
> And we were just talking about those bones and
> a wind, a cold wind blew through the basement
> and blew up all those wall hangings and blew
> around the room, counter-clockwise, and knocked
> stuff off shelves as it went. And you see that,
> and you think, well, there has to be something
> else.*

Are there consequences to ghost stories? Are ghosts
conjured by the re-telling, either oral or written? If not
summoned, does the story energy somehow become
stronger when the narrative is repeated?

I wondered if a shaman could answer me.

Shaman or not, Maynard was wrong about the
Czech; I won't be bringing him. He was wrong about
Sheriff Nottingham and the horrible Indian expression.
He was right about Karl's obsession with his research
work; that was the breakup excuse.

"He's a cancelled Czech now," giggled my friend
Harp about Karl.

July 6th

STORM COMING

The first rule of physics states that energy can be neither created nor destroyed. Karl's love and lust were turning into ache. When the pain re-surfaced, I forced myself to write.

Maybe I shouldn't have the distraction of a relationship either. Maybe love muddies everything. Maybe the freedom of no relationship can free up brain energy to create. Maybe this is what Joseph Campbell meant by organs in conflict.

I wondered why there are no Dorchester ghost stories about people who abandoned love for work or art; certainly Willimina traded love for society. Since mid June, I had written the tour from heartbroken Willimina's point of view; I chose her disconsolate voice. On an unconscious level, maybe I saw my relationship end approaching and denied it.

I suffered a quick and severe fever after the break-up;

in an hour I had chills and 102-degree temperature. It's surprising and amazing the changes in the body when the temperature rises up three degrees: the aches, the dizziness, the exhaustion, and the disorientation. I kept seeing black and white shapes in the corners of the room and under chairs and bookcases.

Hannah and Sarah had fevers, I thought. Willimina died in childbirth; she was delirious. Try to write now. Riddled with fever, I dragged myself up on one elbow and thought: write how this feels. I could barely hold the pen. It was so heavy.

That's fiction or fictionalized reality or folklore or history. We are human and riddled with fever, emotion and subjectivity.

After the cabin fever of the flu fever and still itchy after Karl's departure, I longed to leave town, so I accepted a friend's invitation to weekend in Bethany Beach. I was glad to see the Bay Bridge and to be on the eastern side of the state. I liked being back on the half next to the ocean; its steady breath soothed my heartache. I waved in the vague direction of Cambridge as I drove along Route 404 to Delaware. I felt pulled by the town, by its stories, by the river, but I had my friends Casey and Dana in the car. We were on a schedule; I couldn't suddenly decide to pilgrimage to Cambridge. It was raining lightly, and we passed an accident, the emergency lights flashing on the wet pavement. A car had skidded and landed in a ditch beside the road.

Drivers craned their necks to spy the carnage, fascinated and repelled by death. I had ghosts on the brain. I had heard Casey's ghost stories. I asked Dana if he had any to share.

"Well, I don't know if you'd call it a ghost story," he said slowly, "but I saw something weird in Chicago once." Dana was born in Ohio and raised in Harrisburg, Pennsylvania. He's a secretly smart history buff with a very dry sense of humor. He makes up words, and he's an expert at the movie game.

> *Dana was working in a Gino's Pizza in the east side of Chicago in an old building. Not long before he began working there, a Gino's waiter had died in an automobile accident. During a meeting with his manager in the manager's office, Dana noticed in his periphery that another man was suddenly in the room but hadn't entered through the only door. He was sitting quietly, not far from Dana, but Dana was too scared to look directly at him.*
>
> *"Do you ever get the feeling," the manager muttered, "that someone is sitting right over there, wearing a yellow shirt and a moustache?"*
>
> *"I'm getting that feeling right now," Dana confessed.*
>
> *"Okay," said the manager, "We'll count to three and both look." They did and turned*

*and they were alone. The manager identified the
ghoul as the dead waiter.*

*When he was alive, the dead waiter had
a habit of flicking the backs of his co-workers'
necks. Dana reached over and flicked his fingers
through the hair at the base of my neck as
I drove. "They said that after he died, they
would sometimes feel that on their necks in the
restaurant. By the coffee station."*

"Oh, gross," I said.

"Grimy," said Casey from the back seat.

"I don't know if that was him in the room," Dana
said, back peddling, "but there was no other door into
the room and he was suddenly there and the manager
saw him too."

"We're not alone in this plane of reality," I said.
"Why do we insist on insisting that we are? Fear? Ego?"
The undulating fields of Delaware whirred by, thick with
fat corn. I had the window open and the air conditioner
on so we could breathe in the Eastern Shore.

"Fear," Casey said definitely.

That first night at the beach, Casey and I went for
a walk. The rain clouds had dissipated, and we felt
the need to stand by the ocean. The stars were clear
and as bright as the faux ones in a planetarium. We
searched the sky for the constellation of Pleiades that was
supposed to be very visible that night. We strolled on the

sand, Casey up on the dunes and me by the surf. Casey doesn't like walking on wet sand. A middle-aged couple passed between us, a man and a woman in casual beach attire. The couple looked at me, and I felt the same sensation I felt when the hard ghost man in Fells Point looked into my eyes, like they could see into the ugly basement of my soul.

"Hello," I said, my voice dry. They said nothing.

I thought it vaguely odd that they were wearing no color; they were all in black and white.

The man said pointedly to me, "They're coming back to earth on my birthday." The couple continued calmly walking down the beach. I heard the woman say "this month" and maybe "what" but the wind carried away the rest of the sentence.

"What? Did you say The What?" I yelled into the wind, but they kept walking south.

I ran over and grabbed Casey's arm. "Did you hear that? Back to earth! He said back to earth!" I whispered.

"What?" Casey asked. I turned, and the beach was empty of everything but moonlight. I could see for miles in both directions, but the couple had disappeared.

"Crap. They were just here. I swear. A couple, in their fifties, maybe, and the man said that they were coming back to earth for him. On his birthday." I could see Casey grapple with the belief of my experience. "I'm out of here," I said. We bolted back to the beach house, freaked out, under light raindrops, convinced that I had talked to aliens. We ran over the dunes, down the boardwalk and back to the condo, laughing, panting and holding our sides. The house didn't offer much protection. How could lumber possibly guard us from intelligent alien life forms?

"The man must've meant Pleiades," Casey reasoned as we told the story to Dana. "They could be coming back to earth."

"That sounds like a great answer," I said, searching for any explanation so I could sleep. I didn't want to know; I wasn't ready to know. We all want an overlap of worlds with the hope of some sort of life in the next, but I don't know if I'm ready to know it all. After all, the Other Side didn't think I was ready.

Each time new friends arrived at the beach house that weekend, Dana told them the back to earth story. The tag line for each new entrance became, "Dana, tell the back to earth story!"

That night, I had another getting off a ghost ship at the cemetery gate dream again. This time, as I docked, Christ Church's graveyard was packed with people standing by the graves. At first I thought that they must be the dead, standing by their headstones. Then I realized that they were all the characters from the ghost stories: headless Big Liz; Hannah in her shroud; Harriet Tubman singing; Black Beard with firecrackers in his hair; the old man in his wife's coffin; Bloody Henny with a box of nails; the girl who vomited snakes; the red-haired old lady and dozens of her dogs and cats; the man with his boots filled with gold; the peg leg sailor; and the Scotch brother with his clothes on backwards. I stood on the long boat's middle seat, balancing, hesitating. I didn't want to get out. A blind LeCompte in the first row raised his arm and removed his pair of dark glasses. His eye sockets were hollow and dry and glowed red. They slowly raised their arms. I think I fainted. I remembered falling into the water. I woke up, sweating, curled tightly into the blankets.

I'm supposed to tell these stories, I thought. If it's the last thing I do.

Back in Baltimore after the beach weekend, I resolved to embrace my phantom experiences and to use the diary book as a means of collecting and chronicling my stories, my friends' stories, and the story of the ghost walking tour. I told my friend Noel my plans. He's from Pennsylvania, and he has big brains and little hair. He loves Voltaire and poker. He told me a time expansion story. He once spent a weekend in New Orleans that was too full of memories than was physically possible in that short period of time.

"It was a time expansion or extension," he said. "There's no way I could've done all that stuff in two days. It's just not possible." He smoked a cigar as we sat in lawn chairs in his cement Fells Point backyard. The alley behind him was lit by fuzzy streetlamps, all cloudy in the damp weather. The humidity had rocketed off the scale again. A storm was coming. My knees throbbed. My eyes watered.

"I think the time flexibility thing has to do with all that water," I said, speculating. "Water soaked land is somehow closer to the Veil, whether it's New Orleans or Dorchester County."

"That makes an odd sense," Noel said as scattered raindrops hit our heads. "From a purely physics basis." He stood up. "Was that rain?" He asked, shaking the water off his arms.

I looked down the alley, and he turned, seeing the expression of wonder on my face. The edge of the storm front marched up the alley, a curtain of thick water, a solid wall of rain, striding steadily towards us, enveloping garages, streetlamps, telephone poles and parked cars as it marched. The front drowned everything in its steady path.

"Oh my God," I said, backing towards the house. By the time we reached the kitchen porch, it was teeming. We were drenched. His cigar was soaked. A small lake formed in the cement pad. Our friend Carlos met us at the back door.

"Is it storming again?" He asked, peering out into the back yard.

"It's raining like crazy. Maryland was talking about

water and the Veil, like she does, and a storm came up the alley. It came up the alley like a frigging train. I need another beer," Noel said, pushing past me and heading for the fridge. I dried my glasses on my damp dress.

"Are we playing poker here or not?" A voice complained in the dining room.

"Maryland conjured a storm," Noel muttered, passing me with a beer, back to the gaming table.

"That's not possible," I protested, feeling branded like a witch. "You should've seen it, though. It was the front of the front."

Silence.

"Have you bid yet?"

This is not over, I thought. A beer will not solve this.

A black shape zipped by the poker players' feet. I swung to follow it but said nothing.

David Hunt, the curator at the Smithsonian, left me a voice mail; he needed more information before he could trace the burial artifacts in the Smithsonian database. I tried to send him several emails, but they got kicked back.

I had delivered the tour to Delia, and yet I continued researching half heartedly. The stories have crawled under my skin; that's no great surprise because they're great stories. I tell them to friends to maintain the oral tradition, no matter the possible conjuring danger.

.

July 8th

A HUMAN HEAD

The mutant Dorchester tomatoes developed quickly from hard green bumps to fully fleshed, plump, ripening fruit whereas the Baltimore batch by the fence barely had yellow blossoms. The Dorchester plants multiplied and thickly covered the side wall of the house. The Cambridge plants choked out everything nearby, even the azaleas, and oddly, the honeysuckle vines wrapped around everything in the side garden but the tomato plants. One tomato tendril had wedged itself into a crack in the foundation masonry. Was that crack always there? I could't find it in myself to pull them away from the house. What retribution could they possibly wreck then?

It rained lightly as I weeded the honeysuckle off the forsythia bush. I was filthy, sticky and coated in sweat.

David Hunt from the Smithsonian called me on my cell.

"Did you find anything?" I asked him, trying to focus beyond the heat. A thin film of salty water stuck to my cell phone; it's often gross to be human.

"Did you mean Cambridge, Massachusetts?" David asked, confusing the towns. He was pleasant but baffled.

Even with the right address, he couldn't find any appropriate listings of human remains from Cambridge, Maryland in the Smithsonian physiology database. He found one listing from the banks of the Choptank of a caldarium, a human head sans jaw, donated by Dr. Elmer R. Reynolds in November 1887 to the Army Medical Museum. After the Civil War, army doctors scored promotions in exchange for artifacts, and there were plenty of bones in shallow graves all along the Eastern Seaboard. The Army Medical Museum collected a lot of skeletons and eventually merged into the Smithsonian.

Something about the jaw discovered on the riverbank jarred some click in my head. My backyard tipped. A bird called. Something rustled in the forsythia bush. In my mind, I saw the edge of Cambridge Creek lapping up against the Christ Church graveyard wall like in all my dreams.

Rivers move.

The banks of Cambridge Creek are now in the backyard of the Cambridge courthouse, I thought.

In my side yard, the tomatoes waved in the warm breeze. Several red ones were almost ready to pick. I wondered if I should eat one. A butterfly flew around the tomatoes, avoiding them as carefully as the honeysuckle did.

"I'll convey your request to the curator of the archaeology database, Jim Crocker," David promised. "Still, you might want to check the registration or archive databanks." A day later, he left me another message that Crocker had found nothing about the Dorchester County Courthouse or any artifact submissions from Cambrdge.

"Once the creek went all the way up to the church. Look at the old maps," Maynard had whispered in our interview.

The church is across the street from the courthouse and the creek is behind the courthouse.

When the land was a Native American Indian Reservation was the courthouse ground submerged? Could the What be a water spirit or a sea serpent? The snake in the photo, the water stain snake on my wall, the water stain snake in the Center's upstairs hall, were they all coincidence?

All those dreams I had of water lapping against the graveyard, and I hadn't figured it out.

I was as daft and blind as the *Poltergeist* family. How can we teach ourselves to believe the unbelievable? I wondered.

Faith.

I had to go back to Cambridge.

Two of my Cambridge library books were weeks overdue, but I feared the unmarked graves by the Return Book slot. I feared the marsh that knows no bottom. I considered mailing the books to Ellen.

Back from her vacation, Judy called me. Feeling oddly paranoid, I worried that she was calling about the overdue books. Instead, she gave me the name of an ex-High Street resident, David Orem. "I know you turned in the tour, but you might want to talk to him anyway," she said.

I found that sweet, but she also surprised me. How did she know that I was still obsessed? She too must believe in ghosts and understand their reach.

David grew up in the Josiah Bayly house next to the graveyard wall. "Our house used to sit near where the courthouse is now, across High Street," he said. "It was the oldest house in town and once faced the river. It sat right on the river. They moved it years later. Some say that when they were building it, a workman saw an Indian, a man through a window, standing where there wasn't any flooring." He laughed. "But we never saw any ghosts," he added quickly. "Or the floating Indian."

Houses move on High Street.

Before the jail and the courthouse construction, the Bayly house sat next to an Indian burial ground, in its soggy backyard, between it and Cambridge Creek. The Creek didn't lap against the church land while there was white settlement on High Street. But before the Bayly house and any other white man structures, the Choptank River covered the courthouse lawn all the way to High Street.

In November 1883, Dr. Reynolds found one lone skull on the bank of the Choptank River.

Maynard said look around 1882 when the church burnt down. An architectural history book assigned the jail's completion as the fall of 1884.

It's the ghoul in the pool all over again. Like the banging bed in the Travis' hunting lodge and the basement blood in *Amityville Horror* and the swimming pool in *Poltergeist*, something had to be going on in that courthouse.

17,000 years ago, the ice from the Ice Age began inexplicably to dissolve, and as it melted for 8,000 years, it drastically altered shorelines worldwide. The old pre-Ice Age shorelines are estimated to be twenty miles underwater now. Life is cyclic, and the water is returning. The ice caps are thawing again; a 41-square-mile glacial chunk just snapped free. According to the BBC and NASA, glacier ice off the coast of Greenland has been melting much faster in the past five to ten years, and that speed is affecting climate change. As the Atlantic floods with fresh water from melting ice packs, the thermohaline currents shift, and that swing will throw the planet into dramatic weather patterns. If all that polar ice melts, sea levels could elevate twenty nautical feet.

All of Dorchester County is less than 25 feet above sea level, and it's rapidly submerging. It has lost thirteen of its islands since map-making began in the Chesapeake, and Hooper and Smith Islands are the predicted next victims of the encroaching bay. The Atlantic sea level has risen three feet since 1600, and worldwide temperatures might increase as much as ten degrees by the end of this century. Maryland loses about 260 acres of land to sea per year, at a rate significantly faster than the rest of the world. To make matters worse, the land surrounding the Chesapeake Bay is slowly sinking as the water ascends.

In a conservative forecast, the Bay is reclaiming the land at a rate of a foot per century and with each nautical foot rise, ten feet of land immerses. At a much more realistic two-foot rate, by 2100, half of Blackwater National Wildlife Refuge and most of the southern half of the county will be under water.

The sooner we tell the stories, the better, I thought.

When Dorchester County becomes as submerged as Atlantis, will man be ready to tell a new myth? It might be a myth in which water is the enemy.

July 10th

IT FLOODS A LOT HERE

American history evaded me. It zipped around the base boards and under the bookcases like the evasive black shapes.

My friend Megan's mother worked at the Maryland Historical Society. I called and left her a message about the Dorchester County Courthouse, but she didn't return my call.

I was nagged by the constant presence of the Cambridge mystery calling out to me. I had to go back. I had to go to the courthouse. I had to go alone. I was itching to breathe on the other side of the Bay. I didn't care if the project was finished and the check was in the bank. I didn't care that I shouldn't waste time on non-billable research. I didn't care that I had other project deadlines. I had to find out more about sea serpents.

Just when I was brainstorming on some excuse to drive to the Eastern Shore, Judy called; she sounded excited. She invited me to a trial run of the ghost walking tour. I accepted, relieved that she still didn't know about the overdue books. The tour rehearsal was on Monday, so I spent the weekend in Delaware, visiting Debbie at her mother's shore house.

I felt oddly centered as my car bumped off the Bay Bridge onto the Eastern Shore. My muscles relaxed, and my skin tingled. A patchwork of corn shimmered along meandering lines of trees and the stalwart march of telephone poles. Route 404 was shouldered with badges of black-eyed Susans, daisies and Queen Anne's lace. The smell of earth and cow blew in my window under the tang of far away sea salt; it cleared my head. A faint perfume of decay permeates the Eastern Shore and somehow aligns it closer to the Other Side, right up against the thinnest section of the Veil. Maybe a proximity to nature is the same as a proximity to the supernatural. As Margot Adler described paganism: "all things – from rocks and trees to dreams – were considered to partake of the life force. . . It is a view that divinity is inseparable from nature and that deity is immanent in nature."[79]

My mind felt astoundingly clear.

The power of the universe, the thing we humans call deity, surrounds us in trees, earth, rocks, lakes, and rivers. It shimmers in our cells. It sparkles in light on water. It vibrates on cricket's wings.

Time is bending with space and undulates like the corn in the breeze. I could feel it under the wheels of the car. The planet's spinning on its axis and rolling around the sun while the moon circles us, and I careened along the curve of the earth's rich breast. I looked at the clock in the dashboard and looked away at the fields and looked back and ten minutes were gone. No wonder I'm always late; I was half an hour behind my estimated arrival at the shore house.

A maroon van with *Treasure the Chesapeake* license plates sped up on the linear line of the road and passed me. The treasures of the Chesapeake are its stories, I thought. Not crabs or oysters or clams. I wondered if the Maryland Motor Vehicle Administration would agree.

"The power of tradition and the art of oral storytelling, though diminished by the thrust of mass media, are by no means dead on the Eastern Shore,"[80] wrote folklorist Carey in the 1980s.

Rippling squares of corn rolled to the pine-edged horizon. I hadn't seen a farmhouse for at least two songs on the radio. In the middle of Eastern Shore nowhere, telling stories gives people something to do. Shore tall tales are old realities retold and painted with a tinge of hyperbole, constantly evolving and outliving time and us. Folklore is what we leave behind, the customs of a people. It's too powerful to blot out. We narrate stories to calm our fear of death, and ghost stories do that best. They are almost like little miracles. They all say: your consciousness will not end. There's a life beyond this temporary reality, beyond what we can possibly understand. They are clues to what is real. If people can create reality, then we've created ghost folklore to help explain that reality.

We are a culture of storytellers, all of us: every politician, nurse, mechanic, cook, bishop, and banker. I once thought that my friend Tom, who co-wrote that wacky science fiction play with me, should follow an artist's life as an actor, writer or director. Tom explained that he entertained his family and friends and that was much more important than amusing total strangers. He was right. Families and circles of friends share mutual bonding histories. My family tells and re-tells our summer trip legends. My circle of friends still reports the back-to-earth story. Tomi re-tells the ghost spotlight story. We recount the ghosts in the bathroom stories.

Towns collect stories. The minute town of Trappe holds the prize for the most outrageous myth of the Eastern Shore.

> *Residents claim that in September of*
> *1928, the sky above Trappe rained millions of*
> *tiny frogs, no bigger than a fingernail. At two*
> *o'clock on a clear Sunday afternoon, a sudden*
> *storm whipped up and deposited the frogs in a*
> *few minutes. Witnesses said the cloud looked*
> *like a tornado, funnel-shaped and twisting.*
> *The frogs hopped in one direction, towards*
> *a millpond on the edge of town, resulting*
> *in sickening, ankle-deep piles of writhing*

amphibians. The squirming creatures popped when people stepped on them, and the dead ones stank. The U.S. Weather Bureau blamed a freak waterspout that swirled up millions of frogs from a nearby pond and hurled them on Trappe.[81]

When I arrived at the shore and told Debbie about the Trappe frogs, she nodded and said, "Waterspout." She had heard that Eastern Shore tall tale. I laughed. "Don't laugh," she said. "I've seen a waterspout over the ocean, and they're serious. You can't predict them."

A waterspout is a tornado over water. Tornadoes have a lower rotation over water, so waterspout funnels are more apt to touch down. Every state and every continent except Antarctica has reported tornadoes. 1,000 tornadoes are reported every year in the United States. A tornado is formed when a violently rotating column of air drops out of a super cell thunderstorm. One super cell thunderstorm can drop six tornadoes. Winds at 100 m.p.h. can pick up trucks, and an F3 tornado packs winds of over 200 m.p.h. A 300 m.p.h. twister can pull off shoes and socks and throw debris 200 miles. A tornado funnel pulls a fierce updraft of wind, and a waterspout could easily collect a bunch of frogs from a pond.

"A tornado in Gainesville picked up a seamstress and dropped her a block away, unharmed and covered in mud," I said.

"I think a waterspout could carry a lot of live frogs and drop them on a town," Debbie said.

"An Oklahoma tornado was clocked at over three hundred miles per hour," I said, trying to imagine it. "The noise must be horrible."

I unpacked while Debbie read a first draft of this diary book. "Maybe the Cambridge residents fear the wrath of the ghosts," she said, speculating on all the town denial. "Like the way people fear tornadoes. We know they're out there, but we hope they'll miss us."

"They get flooded a lot," I said, hanging up a dress.

A long parade of people in this experience denied ghosts: Ellen, Thomasine, Martin, Olivia, Lynn, Mrs. Travis and the entire towns of Easton and Cambridge. The Smithsonian found no record of Cambridge bones. I buried my own ghost stories deep under my skin until this narrative shook them out, like stunned bugs from a curled up rug.

Debbie's very sensible; so I didn't tell her about the water mark in my bedroom, the water drop that burned my skin, the tomatoes that cracked the foundation of my house, the pipe noises in my walls, the research hidden in the kitchen cabinet or all the ghost dreams and black shapes. One day more with me, and she might notice that I was turning to water and finding it harder and harder to read. My eyes have a constant film of fluid over them. I don't want to think about it. If Debbie knew, she would tell me to go to a doctor.

I walked down to the beach and stared at the Atlantic, wondering about the Pleiades Man and his birthday. The waves lapped at the ever-changing land, and the air thick with brine centered me. Fat clouds built over the horizon, blocking out Atlantis and Portugal, boiling up big questions in my tiny twelve percent brain. America's not a particularly reflective society. We let others think big thoughts for us. We don't ask big questions on a daily basis. If we think big picture, then that big picture is safely compartmentalized into the comfortable hierarchy and parable of religion. Every American funeral necessitates another review of our repressed concepts of temporality. We separate ourselves from death in order to sanitize it. No one much sees it any more. We no longer wash the corpses and lay them out on the dining room table for three agonizing but edifying days.

A particularly raucous gull snapped at my bare feet.

"What do you want?" I asked stupidly. "What message do you bring for me?

Maybe that's the simple reason why we deny the existence of ghosts. We don't want to die. We turn on the radio in our heads to block them out like the old man photographer in the Dorchester Arts Center. It's too much for our little nervous systems.

All these ruminations by the sea seemed too bulky to me. There were too many words rattling around my head; they made a terrible noise under the sea's crash.

"Did you say hello to the ocean?" Debbie asked, sitting down beside me in the sand. "I brought you a beer," she said.

"Thanks, hon." I popped the top and drank. "Hello, ocean." I shuffled to the surf and poured part of the beer into the water.

"Beach tradition," Debbie said, smiling. We had been pouring small amounts of alcohol into the sea for good luck for years.

"A little something something for Neptune," I said, realizing that we were like the old men dumping fish into the creek and the people who believed that blue boats brought bad luck.

"Crab cakes tonight," she said. "And the fresh corn I picked up on the way."

"And the tomatoes I brought." I wondered if I should be afraid to eat more Cambridge tomatoes grown in my side yard. I had eaten one at home, and it just tasted like a tomato. I drew swirls in the wet sand with my finger. "Do you think people ignore folklore?" I asked.

"People don't ignore it; it's just not fast enough for us," Debbie said, sighing. "Society's so complicated."

"Maybe myths simplify it for us," I wondered. According to the doomsday prophets, mankind's on the evolutionary brink of a new age, a new world, a new reason, and a new myth. "In this transitional period," I speculated, "Maybe the 2000-7000-year-old myths don't always fit fast society."

"That makes sense," she said. "Each generation reinterprets history to fit their experience. I know. I'm a lawyer." She grinned at the Atlantic.

"Maybe local ghosts and folklore are transitional myths. Maybe they can help us through this scary conversion."

"Mary, are you suggesting that we visit Big Liz?" She asked.

"I think I'm going to tomorrow," I sadly said.

"Mary, you'll barely be here a day and that headless girl creeps me out."

"I'll be there during the day and I have to see the places where the stories happened."

Debbie kicked at the sand. "Okay, but call me when you get out," she said.

I don't know anything much, but I know this. We should listen to our folklore, slow or not. It's not the silly, quaint and countrified foolishness that we think it is. It's not random and nor is it far away. For almost thirty years, teenagers have pilgrimaged to the Green Briar Swamp at Halloween to reenact Big Liz's tall tale, and folklore claims that over half those nights have ended in accident, car or personal injury.

The elaboration of legend continues. There are still blind LeComptes. There are ghosts. There are monsters. In his 1964 compilation of Baltimore ghost tales, Karl B. Knust told the horrible story of two vampires buried at a crossroads off Interstate 83. The thin hole in the Veil is closer than you think. It's in suburban Hunt Valley. It's under your house. It's in your backyard, like the Baylys, like the woman with the hair bird nests, like the mutant tomatoes tearing at my foundation.

We ate the mutant backyard Cambridge tomatoes with the crab cakes and the corn at dinner. They tickled my throat going down, but tomatoes are acidic.

"These tomatoes are very sweet, Mary," Debbie said.

"They're from Cambridge," I confessed and told her the story.

"Well, they might be mutants, but they taste fine to me," she said. "They probably grew at different speeds because of some soil difference between the two gardens."

"I put the same fertilizer in both," I muttered, but I hadn't considered a soil difference.

The next morning, I left Debbie and drove to Dorchester County from Bethany Beach along Route 26. I passed the Scuba Bible Adventure School and the Inner Peace Bonsai Studio. A storm followed me inland to a copse of trees at the Maryland border; its white clouds arched against the slate gray sky. When I drove out of the other side of the trees, the storm had vanished, the sun was shining, and it shone all the way to Dorchester County. I snaked down country lanes, searching for DeCoursey Bridge. No litter on these back roads; the natives love what little land they have.

Dorchester County thoroughfares have modest signage except for the no trespassing signs. The natives don't want the world to find their corner of wet paradise, and they don't need directionals. The natives still drive El Caminos, and they drive fast. Their world is so isolated that three old men waved to me from their meticulous front yards. There are no straight highways in Dorchester County; they are riddled with turns and yet I never lost my sense of direction. I seemed to know intrinsically where I was heading. One turn and the road fell away and water rose up, seeming higher than the tarmac with only hassocks of swamp grass keeping the river from my car. Islands of pine climbed out of the creeks. Eagles and hawks rode the air currents like surfers. The marsh water seemed shallow; dozens of chair-sized atolls formed the spine of a sleeping giant in the waves. Standing in the salty breeze of Blackwater, I understood the story of Molly Horn and how people can imagine stories from geography. The light that reflected off the river held secrets; maybe that watery giant was the devil. I shivered in the heat. There's no pull over on the side of these two-lane paths. One wrong turn and into the muck sink you. I understood the Wallace marsh story; the county roads were built on shifting landfills rising up and out of the sinkhole of the swamp. Pine trees towered over the telephone poles. I understood the Henrys Crossroads ghost-in-the-tree story.

I spent a half an hour talking to the senior volunteer

in the gift shop of the Blackwater Visitor Center. We
were the only ones in the center; she seemed glad for
the company. I told her that I once slogged through the
refuge's swamp as a teenager.

"Oh, you moved to the big city," she said, nodding
her white head. She wore a green apron over her blouse
and khakis.

"Oh, no, I was born there. In Baltimore."

"Oh. I see." She seemed sorry for me.

I told her that I wrote the ghost walking tour for
the Arts Council. She changed the subject to discuss
woodland hiking paths. I forgot to ask her about the
green-eyed mule. She might not know anyway; she
lived in Easton. She gave me a map of the refuge; all
Dorchester County maps were slightly different. Some
tracked roads that others completely omitted.

I ventured out again and discovered that DeCoursey
Bridge Road doesn't intersect Route 50; it ends in
Drawbridge Road that passes Austin Road. The Austin
plantation wasn't far from the Green Brier Swamp; I
wondered if Big Liz and John walked or took the wagon.

Roads changed names in Dorchester. Churches
and houses teetered up on blocks. Cattails grew in the
drainage ditches where wild turkeys and quail hid. Red-
winged blackbirds flocked. Herons balanced on one
bumpy leg. Dragonflies nudged the car windows. Deer
quivered by the forest edge. The road bent and in its
arm snuggled a bumpy slave graveyard or the elbow of

a creek decorated with water lilies or a perfectly round, bright green slime pond. One turn in the woods and I drove by the Breeze Away Beauty Salon in the middle of nothing but fields for almost twenty minutes. Another turn and a crooked No Hunting sign leaned in front of an abandoned meeting house. Another turn and shirtless fishermen swung line out into the brackish mix of salt and fresh water. Pine forests surrounded fields of corn. Suddenly, with no break, the corn turned to cattails and thicket bushes, the pine woods thinned out to lonely waist-deep sentinels, and the Green Briar Swamp spread out under the sky. DeCoursey Bridge was a low concrete bridge, spanning over tea-colored water that was almost level with the bottom of its knee-high guardrail. Natural debris gathered around the bridge's pylons, choking the flow of the river to the bay. The swamp grass was mottled like an Impressionist painting; bright green blended into mustard yellow and burnt brown. The sky hugged huge meadows of grass dotted with pockets of dark water. A white couple in their forties was crabbing off the bridge, pulling up a wet trap to check for bounty. I parked next to their red truck, and they dubiously watched me. My foot hit the bridge, and the wind whistled through my long hair. I introduced myself.

"Isn't this Big Liz's bridge?" I asked.

The man was surprised. He was tan and blonde. "Well, yes, yes, it is. I thought that was just a story my mother told us," he said, smiling. "To scare us away from here."

I stood on the bridge and told Bruce and his wife Annette the story of Big Liz.

"Have you been here at night?" Annette asked. She taught middle school in Harford County.

"Are you kidding?"

Bruce laughed. "I wouldn't come here at night. I wouldn't do it before and now I've heard you talk, I never will. My folks bought property here in 1960, down in Bishops Head, but you gotta understand that I'm not a native."

"I get it," I said. He was born here but not a native.

"The natives are close-lipped," he said. "Some old guy showed me an arrowhead when I was a kid but he wouldn't tell me where he found it."

"You should come here at night," said Annette. The sun had dappled her nose with freckles.

"There's no way," I replied. "There's an angry slave girl out here."

"Did you know that Bishops Head's called that because some bishop lost his head there?" Bruce asked.

"I had heard that. Lots of headless ghost stories in this county." I told them the story of the headless rider on the Talbot shore.

"I'll tell my kids about that one," promised Annette. "It's a great way to get them to think about the American Revolution."

"That's an awesome idea," I said, pleased that Harford County students would learn Dorchester folklore.

"Thanks for the story," Bruce said, shielding his eyes from the glare of the sun on water. "I'll tell my Mom she's justified." We laughed more, and I walked back to my car. They returned to their traps.

As I drove away, my cell rang. I fumbled for it and accidently hit the horn just as my wheels hit the bridge. I giggled nervously, and Annette looked startled. I had finally found the bridge and summoned Big Liz, but she didn't show up with her detached head.

I drove down to Bestpitch, the town where the two old people disappeared, leaving a dead pet behind. Bestpitch is completely surrounded by swamp. To access it, you drive up and over a one-lane, restricted, railroad-tied bridge over the Transquaking River. As far as I could tell Bestpitch is comprised of only three houses. As I drove up to the houses in the middle of the marsh, a hulking, black dog loped out into the center of the narrow highway. I slowed the car and he watched me, drooling. Finally, he sprung into the drainage ditch and let me pass. I felt oddly cold, although the sun was shining. Creeped out, I drove on to Cambridge.

The sun shone on Cambridge too, yet the street and sidewalks were damp. I found that odd, since I had not passed through another storm. I pulled into the library parking lot; I had to return my borrowed books

or end up in the graves with the Woolfords. Hoping
not to see Ellen, I dropped the books on the main desk
and asked the location of the Maryland Room. The
Maryland Room of the Cambridge Library is a ten
foot square corner of the main room. A librarian
showed me the shelf in there where I might find some
books about Native Americans. I didn't ask her if she
was Thomasine, but her voice sounded deeper than
Thomaine's high-pitched screech.

"I never heard about an Indian burial ground,
and we don't have any information on the courthouse
being built but this shelf has some books on Indians
and pirates," the librarian said and left me. Clearly,
the Enoch Pratt librarian had been wrong about
the Cambridge Library Maryland Room. Clearly
Thomasine's co-workers realized that pirates existed.

I researched for several hours, killing time before
my arts council meeting. A storyteller read a book to
a group of riveted children by the main desk. An old
woman looked through the catalogue station in the
Maryland Room and nodded sagely to me. I found
a few more facts about Indian religion and a book
published by the Maryland Historical Society about the
state courthouses. "According to Elias Jones, the first
courts of Dorchester County were held in homes."[82]
Somewhere around 1687, a courthouse was built within
the town of Cambridge, although the book claims its
land grant was on the Transquaking, the meandering

river down by Bestpitch, and not the Choptank. That
17th century courthouse, wherever it was, was too small
for its burgeoning community, so it was replaced in 1797
by a brick one on an adjoining lot. That brick one may
or may not have been built directly up against the home
of Joseph Dowson; a name I didn't recognize. In 1852,
that brick courthouse burnt to the ground along with
all the records and papers of the court. The third and
current courthouse was built somewhere around 1854.
It's dated then because it was mentioned in a meeting
that year, involving Governor Thomas H. Hicks. Not
until 1931 was that courthouse scheduled for "repairs,
renovation and enlargement."[83]

Except for the 1852 fire, none of this documentation
fit my previous research. It didn't jibe with Winston,
Maynard or Father Martin. Who should I believe? I
wondered, my head in my hands.

Resigned to the close-lipped residents and the
contradictory reports of history, I returned the books
to their shelves and marched determinedly out of the
library and past the courthouse, longing still to visit the
Register of Wills, but scared of the building's watery
secrets. I walked past the Neighborhood Crime Watch
sign. I checked the windows in #116 High Street for
any movement but there was none. Everything seemed
smaller in scale somehow, like a model, like a movie set.
I willed myself to enter the Dorchester Arts Center. I
slipped into a full room in the tiny back kitchen where

two, older, Episcopal women sat with Judy and Delia. I handed Judy the empty tomato bucket.

"Thanks," Judy said, standing and taking the bucket. "How are they doing?"

"Oh, they're flourishing. I had some yesterday. I should've brought you some."

"Don't worry," Judy said, smiling. "We've got plenty. So many that we can't give them away. But enough about tomatoes. These are two more of our Arts Center board members. Rickie and Doris, this is our writer, Maryland."

I thought these board members were taking the tour, but they only wanted to talk to me, to assess me somehow. They looked like they should be running the Nearly New shop at St. Mary's Outreach Center. They wore bright colors and prints and carried several mysteriously shaped bags. One of them clutched a Harrods department store bag, so we discussed London. We laughed about British dental hygiene and plumbing. They chatted with me for a few minutes and suddenly picked up the bulging bags and left.

Neighborhood watch indeed, I thought.

A volunteer librarian named Jane arrived; she was contracted by the Dorchester Arts Center as one of the tour guides. She was vivacious and bubbly, in her early thirties and slight with straight, auburn hair. She was wearing a flowing jean skirt and a indigo tee shirt that matched her eyes. She carried a thin binder with the

tour text. She settled next to me at the kitchen table.
"So, you wrote the tour?"

"I collected the stories," I said, gently correcting her.

"Do you mind if I ask you a question?"

"Please do."

"I'm worried about the local historical accuracy of one of the tour stories." Jane was referring to a ghost story from Glasgow, Maryland of spectral members of the Constitutional Congress pacing through the home of William Vans Murray. People have reported seeing them late at night, arguing.

"Phantom Congressmen," I said to Jane, pitching the ghost story. "How vexing they'd be. They'd never shut up. Forever. Imagine that."

"Oh, I like the story, and I like that image," Jane said very seriously, "But the house in Glasgow's not William Vans Murray's birthplace." The Vans Murray birthplace was Jane's only correction to the tour script. She didn't challenge the ghost story itself; phantasm senators didn't bother her. She was concerned that locals might recognize that juxtaposition with history, like Joan's reaction to the Bertha's ghost stories in the Fells Point tour.

"I somehow doubt that many Cambridge locals will take the tour," I started.

"Well, yes, me too."

"Still, if it bothers you and despite the romance of dead Constitutional framers roaming the Glasgow house," I said, "We can cut the story."

Jane was relieved. "Oh, thank goodness," she breathed. "I was so worried. Although I do like the idea of them still fighting as they paced." Jane nodded, considering the cantankerous spirits. "What was your source for the Glasgow story?"

"Flowers' book was my source for that," I said slowly, watching Judy and Delia leave the room.

"Well, he was wrong," she said, rolling her eyes.

Strange, because the folklore book has a photograph of a historic marker, stating that Vans Murray was born in Glasgow. Is the marker a lie? Was Flowers confused? Who does one believe? I wondered. When it comes to folklore, nobody is right and nobody is wrong.

"So, are you a native?" I asked Jane.

"Oh, gracious, no, I moved here when I was nine," she said, tucking some hair behind her ear.

"How was that?" I asked, understanding why I felt connected to her.

"Hard sometimes. When I was a kid, I heard Big Liz's story but with a different spin."

"Yeah?' I asked, intrigued. "How wonderful, I mean literally, how full of wonder. I was just at the bridge." I told her about Bruce and Annette.

"I wouldn't go there at night either! I think some kids told me her story to scare me," she said, "and it worked. I never went. I heard that when she was finished burying the gold and she was patting down the dirt of the treasure hole, John Austin asked her, 'Would you know the way back into the swamp? Would others?' Isn't that a terrifying question?" Jane's eyes widened.

She's a natural storyteller, I thought. And storytellers control reality by creating it. "She's dead no matter what she says," I said.

"Exactly," said Jane, nodding, her hair swinging in assent. "No matter what."

I was reminded of my potential visit to Maynard's house on Elliott Island. Czech or no Czech, I'm probably headed for marsh bottom no matter what I say or do.

"Let's put that question into the Big Liz story," I said to Jane.

She looked pleased. "I added to a story from my childhood," she said, running her small hands across the black binder. "How lovely."

Gray clouds crossed the sky in the kitchen window. A single raindrop landed on one of the thick panes. I was so glad that Jane would tell the stories and keep them alive.

Delia led a couple in their early sixties into the kitchen. When she introduced us, they stared at me woodenly. Dan worked in the marina. His wife Beverly snuck hostile glances at me. Jane tried to drum up some small talk to no avail. I felt uncomfortable in such a small room, so I wandered out into the gift shop.

"Judy," called Delia, "Can you join us, please?"

Bored and dying to get to the courthouse, I flipped through the beach pastels, sealed in plastic. I felt sad for the seagulls, trapped under the cellophane.

I overheard Dan protest to Delia and Judy, "The tour isn't dramatic enough!"

Isn't dramatic enough, I thought. Good God, man, there's a sea serpent hulking under the marina! There's a headless swamp girl guarding treasure! The Indians are making live fish sacrifices! The whole town's in denial! There was a naval battle at the end of the street!

Judy urgently gestured me back into the kitchen. I hesitated because I didn't want to be in the same tight room with that unhappy man. The Center wanted to contract Dan as a guide because he had community theatre experience, but he didn't like the script.

"It's not a script! It's a sixteen page monologue!" He cried. He wanted actors jumping out from behind bushes and portraying every ghost, like the Hampton Elementary School haunted house tour.

I never wanted to be part of this town's arguments, but I obviously had no choice. "The Center clearly does not have the resources to pay stage actors to leap out of bushes," I said with no irony even though I once requested a drenched guide. "Please try to understand that." Dan crossed his arms and I continued. "I'm writing this tour in the storytelling style, in the history of 400-year-old oral tradition, the oral history that told these stories. It's just a genre difference, and barely even that." There was a long pause in the tense room. The wife Beverly looked confused. Judy played with the edge of her pale yellow sweater and stared at the floor. Delia's face had turned to granite. Jane tried not to smile. Dan's tanned features turned beet red. I carefully continued. "I guess you see a broader difference than I do between actor and storyteller."

Dan shook his fist at me. "I know actors! Shakespeare wrote for the stage, not to be read," he said.

"Well, in Shakespeare's lifetime," I explained, "Quartos, the bound and butchered scenes from his plays, were sold, usually during the play's first run. The library stands at St. Paul's were full of them, and a few years ago, I almost bought one on eBay. The quartos are the basis of several of the texts that we use today, including *Hamlet*. We have no true idea who wrote those stories. Or who wrote any story for that matter." There was another awkward gap of time. Jane looked out the window, fascinated by a tree and its moving branches. The tiniest of smiles cracked Delia's stone lips.

Dan waved his copy of the tour at me, its white pages flapping. There was something yellow on nearly every page. He had highlighted the history in the ghost walk that overlapped with the history in the West End tour. "This tour's got some of the same stuff in it!" He accused.

I smiled. "Well, I'm hardly surprised. It's a similar history of the same town. The West End tour's only a block away. One should hope it's got some continuity," I said. "Continuity brings value and validity to history," I reasoned.

"Maryland's contacted all her sources, and the Center has copyrighted the walk," Judy said, thinking she was defending me.

"I don't care, and I won't do it," Dan announced and pulled his wife's flowered sleeve. "Come on, Beverly!" He threw the script at me, and they stormed out, slamming the front door behind them. The bell rang into a splendid silence.

I announced into the hole they left behind, "The stories are not proprietary, and they do not belong to anyone individually. They don't belong to the folklorist who compiles them. They don't belong to the Dorchester Arts Center. They belong to the state. They belong to the county. They belong to the people. They identify and define the people no differently than the marsh grass and the cattails and the herons." Delia and Judy looked vaguely bewildered. Jane grinned at her script. I had such calm conviction in my voice that I surprised myself. "And we should give the stories back to the community. It's our responsibility," I said. "And I didn't contact all my sources; a lot of my sources were published folklorists because no one in this town acknowledges the fact that we will die and turn into ghost. And I fabricated one of those ghost stories and you know it. You told me to."

Judy and Delia's eyes widened further, but there was no rebuttal. I didn't mention that I wanted to dredge the Choptank or organize a diving expedition under the marina or raise the funds for a flotilla that could comb the riverbed with sonar. I didn't say that I had to find out what was hiding under those boats or it would find me out.

In the gift shop, the cashier rang up a sale. The floorboards above us creaked. Pipes shuddered in the walls. I waited for a phantom hand to cup my human waist. I waited for phantom crying.

"Well, you're right," Delia said in a jovial voice, "You're right about all of it. Let's go, ladies. Let's test the tour. Let's hear those stories that belong to the state of Maryland and the county of Dorchester and not to us." That comment dissolved the room's tension, and everybody but me giggled.

I couldn't tell if Delia was repressing the conflict or embracing the positive. Either way; good for her.

"That Shakespeare stuff was great," Jane said to me as we filed outside.

"It's true," I said. "Listen, Jane, I swear to you that all of the ghost stories except one are from Dorchester County."

"I believe you," she said. "Did Judy and Delia really ask you to make stories up?" She whispered. "Which one? I can't tell."

Judy joined us before I could respond. "Don't worry, Judy," I said. I patted her arm. "I'm sure you can find other actors to do the tour. Maybe there're some up in Easton."

"Oh, to heck with him," Delia blithely announced, as she walked down the Center's steps. "We don't need him!" The Center door closed behind her. The bell rang again in the quiet humidity of the afternoon.

"Maryland didn't want a male tour guide anyway," Judy said.

"I wanted one or the other," I muttered under my breath.

"We have Jane," Delia said cheerfully.

Through the July damp, Judy, Delia and I leisurely ambled the path of the tour while Jane read the script. Jane paraphrased because she didn't have the time to memorize sixteen pages of monologue. I didn't care; that's how legends grow and become the storyteller's myth. I had passed them on to Jane, like Mr. Travis and Maynard passed stories on to me, like other natives had passed stories on to Tom and George and Vernon. I listened to Jane elaborate on her own and thought, maybe that's part of the Other Side lesson. Let the stories go.

Repressed or not, encouraging fabrication or not, Judy and Delia delivered Dorchester folklore to its community, and that was a beautiful gift and worthy of an arts center.

We slowly strolled down High Street, dodging

random raindrops and listening to the underpinning of the county: its legends. The blue of the sky was slowly replaced by gray, bleak clouds that pushed the azure down to the horizon. The wind whipped out our skirts. We leaned on the graveyard wall, watching birds swoop out of Ann Weller's mausoleum tree. The leaves turned backwards.

"It feels like rain," I said.

"They sure do keep the graveyard neat," Delia said, admiring the finely shorn, cemetery grass.

"One line of thinking goes that a clean, well-tended graveyard is ghost-free," I said, watching the leaves dance. "Energy can move more freely through a clean space." Christ Church's yard was immaculate; maybe that's why Olivia and Martin have no ghost stories to tell. Yet, across the street, the sacred bones of Native Americans were crated up and shipped to a crowded warehouse in Washington, D.C., like the final moment of *Raiders of the Lost Ark*.

Debate still thrives about the moral validity of religious artifacts on display in museums. Museum mummies should stay buried; there's something wrong about them trapped in a glass case, surrounded by a sea of tourists. The National Historic Trust in this country has on its list of protected sites a Zuni burial ground in New Mexico, because the Salt River Project power company plowed through human remains in a sacred Zuni salt lake.

Graves have been disturbed in Cambridge too. Human graves are still trapped under High Street; the Christ Church graveyard boundaries have shifted over the centuries. Jane leaned against the graveyard brick wall and read a story from the tour of a burial under local sidewalks.

> *A few years ago, on Academy Street just south of High Street in Cambridge, local children reported playing with a gray cat that no parent ever saw. The parents believed that the children were imagining the story of an invisible cat until a repair of the Academy Street sidewalk revealed a cat skeleton bricked up inside of it. Once the cat bones had been re-buried away from the confining sidewalk, the mysterious ghost cat sightings ceased.*

Human civilizations are defined by how we bury our dead. 5,000 years before the birth of Christ, our ancestors marked their graves. 3,200 years ago, the Egyptians began embalming their dead. Twenty-four centuries ago, in his classic play *Antigone*, Greek philosopher and playwright Sophocles warned his society that divine law is more powerful than man's earthly decrees. We must bury with dignity. If the character of Creon in *Antigone* is wrong in forcing the body of soldier Polynices to lie unburied and fester where he fell in battle, then the white pot-diggers, including cranium-digger Reynolds, should've let the Native American graveyard sleep by the banks of Cambridge Creek. Obviously, the pot-diggers had never read *Antigone,* otherwise they would've known not to remove any dead men's bones.

As Judy, Jane, Delia and I crossed High Street from the graveyard, Jane said, "My favorite tour story's the mystery of the What under the Cambridge Jail. My nine-year-old son loves that one. It's so spooky," she said, grinning.

"Yeah, that's my favorite too. I wish I could find out more about it," I replied.

"There are enough stories in the tour," Delia said.

Truth will out "as liberal as the north."[84] You can't keep history a secret, not with humans and legend around. We stood on the lush, blood-soaked lawn of Spring Valley in front of the courthouse. The fountain tinkled. A Cambridge resident delicately walked a pit bull around the green knoll, clearly keeping the animal off the grass. The dog growled deep in its chest. Its leash strained.

"Dogs freak out at Spring Valley. People can't walk them through the grassy knoll," Delia said. "They have to walk around."

"Why didn't you tell me that before?" I asked.

"I didn't think of it."

Animals lack human repression and clearly sense the supernatural. The vestigial slave auction and witch hanging energy was drenched in its soil.

"I'd love to add that story to the tour," I said.

"I can add it in," Jane said.

Bless Jane; she'll maintain the oral tradition, I thought.

"We could hire someone to walk a dog through the tour," I joked.

"You're very funny," teased Delia.

"My favorite story's the Hurricane Hazel story about the old man whose his wife shows up in her coffin," said Judy. She looked at me intently with her pale blue eyes. "You should see it here when it storms. It's very impressive. I've heard that someone dies in every big storm, usually some older person. In most storms, someone's born, too," she said softly, looking across the street towards the graveyard. A moist wind blew the branches on the tree above us.

"That's an old wives' tale," said Delia, playfully nudging Judy in the shoulder. "We're just tired of mopping our basements and cleaning up after. It floods a lot here. It's very wet."

"My basement's practically a tidal basin," Judy complained. "Three floods in that many years. We just don't keep anything down there any more."

"We get flash flood watches weekly," said Jane.

Delia and Judy bit their lips and turned into the wind. With that, the sky tossed a steady line of raindrops down upon us.

Face facts, I thought. There's something down right dangerous about water here. Embrace the archetype, the daily vitamins of deep principle.

As darkness fell with the light rain, Jane sped up the end of the tour, tearing through the story of the Confederate LeCompte ghost in the house next to the courthouse. She slammed shut her binder. "Okay, that's it," she said. We called our goodbyes and ran to our cars to escape the wet.

I drove behind the High Street houses to turn the car around. Cambridge Creek runs from the river past a seafood company, past the wealthy backyards, past Snapper Cove Restaurant and ends against a retaining wall at the base of the courthouse lawn. Wedged between the seafood company and several backyards was a weedy rectangle of ragged land raised up like an old landfill. It seemed mysteriously grave-like as it spanned the land between the classes. What was under it, I wondered.

I drove past the closed courthouse. Its tall metal doors mocked me with the secrets of High Street in the rain. I sighed, turning the corner and heading back to the Western Shore.

About a half hour later, as I drove onto the edge of the Chesapeake Bay Bridge, my eyes suddenly filled with water and teared up. I tried to blink to see. I hate bridges; I'm afraid that I'll careen off one. A tractor-trailer rumbled beside me. In front of me, a white van slowly trundled up the bridge. The lettering said something about pool cleaning. I down shifted and blinked and reached for a box of tissues in a panic. My eyes stung. I couldn't see. The Bay grew further and further away as the bridge raised me up over the water. My breath became shallow. My chest tightened. I could barely see the lane markers and tried to focus on the back of the white van. I blotted my eyes with a tissue and almost knocked my glasses off.

This is how I'm going to die, I thought. From water, from tears.

Somehow I managed to stay in my lane. Somehow I managed to fight the temptation to drive into the truck or over the guardrail. Somehow I managed to reach the other side with my face soaked and my nose running. Once I cleared the tollbooths, I pulled over, sobbing hysterically.

The stories followed me. They eked into my skin like shower water, and I was crying them out.

July 11th

WE NEED A CARCASS

"I'm back on the Western Shore," I told my friend Adrienne over the phone the next morning.

"Where were you?" She asked.

"On the Eastern Shore," I responded.

"Ah, of course."

"I can come over later," I said, "but first I have to go to the Enoch Pratt for a while."

Every year, new species are discovered. In 1896, after centuries of legend and no carcass, the body of a giant octopus washed up on shore in St. Augustine, FL to prove all those sailors' tall tales of eight-armed monsters. Everyone was astounded, including Jules Vern who had written attack octopuses into *Twenty Thousand Leagues Under the Sea* in 1870, twenty-six years before.

Only an estimated one percent of all fossil bones have been unearthed. The first African gorilla was not categorized until 1915. The Native Americans have over one hundred words to describe the American myth that is Bigfoot. Eight foot tall skunk apes have been reported in Florida for 200 years, especially after hurricanes flush them out of the subtropical Everglades. Tales of fire-breathing dragons are riddled throughout Japanese, Chinese and Mayan folklore. Most of the dragon myth is told nearby swamp where ethane gas can spontaneously ignite with phosphorous or potassium and where methane and oxygen abound. Sea monsters haunt the icy waters from Lake Champlain to the Great Northeast Sea to Loch Ness. Two thirds of the planet is covered in water. The ocean is vast.

The problem with proving sea serpents is that scientists need a carcass. Cryptozoologists, who study hidden, extinct and/or mystic animals, speculate that the real culprit behind these sea monster stories are either an aquatic dinosaur or plesiosaur or a primitive whale called a Zebulon. Theoretically, the plesiosaur died off 100 million years ago, still in 1938 a primitive fish that was thought extinct for sixty million years was caught live off the coast of South Africa.

The first sea serpent account in the New World was in 1639 in Cape Ann, Massachusetts. The Native Americans of Cape Ann believed that a water spirit, half god, half demon, lived under the harbor. Each time they crossed the bay, they threw overboard a sacrifice of fish. Legend says that one day, a fisherman forgot to make the sacrifice and the next time he crossed Cape Ann, he drowned. Locals say that a long, snake-like beast rose under his boat and capsized it.

I had returned to the Enoch Pratt Free Library, this time to research sea monsters. I was afraid to go to The National Archives; I might run into Karl. I put down the folklore book and stared out at the sky, remembering the men dumping fish into Cambridge Creek. The clouds were darkening again. It had been such a wet spring and stormy summer. A bee-sized black shape shot along the edge of the library window. I rubbed my watery eyes and returned to reading.

We've lost the knowledge that our species used to cherish.

I dreamt two nights in a row of rotting floors in my house. In the first dream, I found a huge hole in my living room in the middle of a party.

"How did that get there?" was my reaction to seeing the basement below.

In the second dream, again I found a hole in my house's floors, but this time I wondered how I would finance the repairs. The floor archetype was probably a message from my brain that I am considering my foundations, my basic elements, my belief systems. In searching for stability, I found gaping holes. Maybe in all my considerations of religion, history and myth, I'm finding big gaps in my belief system framework.

Maybe the holes are the holes in my knowledge, lost

like Atlantis.

We know so little: four percent of all reality with twelve percent of our struggling lizard brains.

July 12th

ON MY WAY THERE

I dreamt Maynard and I were sitting on a horse that walked slowly across a meadow thick with goldenrod and flies. The field looked like Gettysburg; it was summer and hot. The horse rocked beneath us as we sauntered past a split rail fence. I thought I could hear buckshot and the sound of distant battle beyond the trees. I wished I could see his face, but he was in front of me. I reached into a drawer in my right arm like a Salvador Dali girl and pulled out an envelope. I knew it had a check in it. I heard his voice say in my head, "Bring the check." I woke up, crying.

I sensed something coming, heading towards me. The barometric pressure felt lower, and it made my skin itch. My sense receptors sent exaggerated messages to my brain. Objects had harder edges; the dresser edge seemed dangerous. The drone of a neighboring lawn mower sounded as loud as if it was in my backyard. The smell of cut grass was dizzying and under it I could almost taste the decay of leaves in the flowerbeds. I felt an odd claustrophobia from living in the middle of so many people. I picked one of the Dorchester tomatoes and ate it for lunch. It gurgled down my gullet and tickled my esophagus, sharp and acidic.

My friend Megan, whose mother worked for the Maryland Historical Society, called me. "My mother never got your message about the courthouse," she said. "The Historical Society has a communal voice mail, and somehow yours was deleted. She suggested that you go to the Maryland State Archives." I searched the Maryland State Archives database online and found no listing for the Dorchester County Courthouse.

I cannot ignore Maynard's Vision Quest for many more days and nights, I thought.

I felt compelled to spread this crazy myth; shared stories shape all of us. This local folklore molds us on an unconscious level; I had grown from Dorchester County archetypes. I was thinking in definitions and patterns. I was thinking how our geography influences our accents and our types of local myth. I was fascinated by the physics of weather. The weather report called for rain and isolated, severe thunderstorms. A cold front was hitting warm moist air, and the Jet Stream was splitting in half over the Eastern Seaboard.

I found myself staring at my computer screen and

thinking about accents and stories. How did I get upstairs?

A land of sinking mud and submerged graves makes for some tall tales. As Christopher Weeks wrote in his architectural history of the county, "Dorchester's miles of shoreline and acres of swamp and forest, with their abundant wildlife, have controlled writer's imaginations, even as they have controlled and shaped the county's development and the countians' minds."[86]

The land under our feet shapes our voices. Dorchester County lies between two rivers and holds the convergence of five others. High Street lies between a creek and a river. The county's riddled with shallow graves, standing water and bogs. Life and death are closely intertwined in a swamp. It changes a body when the only solid footing underneath is isolated clumps of marsh grass. Like ill-fated Wallace, one wrong step off a reed island and you drown in gritty, black mud. The evolution of man has been a blink in the history of this planet, and we're not that far from that dark swamp of our dank origin.

My speech consultant friend Betty Ann told me "many linguists feel there is a strong relationship between our voices and the land." She's from Texas and prickly, funny and blonde. She's never been to the Maryland shore. According to Betty Ann, several of the variables that impact the development of local language are: what language and accent did the immigrating people have upon arrival; was the geography and weather of an area similar or dissimilar to their native land; how did their language and legend integrate with natives and prior immigrant groups; and how did the geography encourage the keeping of their accent or the changing of it.

Same with language; same with stories. Certain lands at certain times tell certain stories. Like a convergence of five rivers, the voice and stories of the aristocracy, deserters, criminals, Roman Catholics and watermen mixed together in Dorchester County to create the likes of Big Liz, ghost ships, Wallace in the marsh, Black Beard and the Hurricane Hazel old man and his floating wife. Damp folklore reflects the countanians' lives. Only myth that was reflective of the water-riddled community was told and re-told, and thirty-two percent of Dorchester County is flat, poorly drained land that will be underwater in a hundred years.

Like a convergence of five rivers were my thoughts.

I shut down my computer and picked up the phone.

I can't stand not knowing anymore, I thought. I can't wait for Kevin the map man or David at the Smithsonian. I can't rely on Judy or Delia or Dorothy for The What answer.

I cancelled an afternoon meeting. As I walked to my garage, my stop-no-more-ghosts friend Joe called my cell. "What you doing?" He asked.

"I'm heading to Cambridge, to do some research."

"I thought you finished that . . ."

"I was. I am."

"Oh, well," he said, "I have to pick up some sconces in Easton. I wasn't going to do it until this weekend but my afternoon's opened up but I can't go until much later in the afternoon."

I looked at the horizon lowering over the bungalow rooftops. It was turning to the color of wet slate and creeping towards the sun. "I better go alone," I said. Either it gets me or I get it, I thought, but I said, "It looks like it might rain."

"Good luck dodging the weather," he said. "Call me later in the week."

The drive from Baltimore to Annapolis was a windy one; the car rocked slightly all the way down. Bumpy gray clouds hung like the underbelly of an old mattress. The leaves in the lush forest along Route 97 turned downhill towards the Bay. Raindrops scattered on the windshield. My knees hurt to shift. My knees are pretty good barometers for storms, and they were both throbbing like the heart in *Telltale Heart:* boom, boom, boom, boom.

Circumventing Annapolis on Route 50, I got trapped in front of a long funeral procession in a traffic jam. The car directly behind me was in front of the hearse and leading the procession. It honked at me several times, its flashers blinking in tandem with the line cars behind it, like the summoning recipe for Big Liz.

I told myself that the driver was wracked with grief. They're dealing with the end of a story, I thought.

I looked down to change the radio station. When I looked back up, the procession had vanished and was replaced by a rather large seafood distributor truck.

Where did the procession go? Did they exit? I wondered.

I wasn't near an exit. I checked my rearview mirrors, and the cartoon crab on the seafood truck mocked me, smiling stupidly.

Not everything means something, I thought.

Then I wondered if I had jumped realities. Was it that quick?

Suddenly I realized that my bra felt damp. Clothes kept dampening against my skin. I was melting. I wondered if I should check my underwear in a courthouse bathroom. I wondered if it would be the second floor bathroom that exploded in 1970.

On the other side of the Bridge, a thin arm of wiggling light broke through the fat cloud waffle and soon blue was mixed with it, aqua patches quilting the sky. The storm took a right and headed south down the Bay, dragging tendrils behind it. It slunk along the southern horizon to a narrow band of dark as I drove along Route 50. When I sped out of the thick copse of trees after the outlet stores, the sun shone flatly on the waving fields. Something, maybe the shadow of a cloud, moved through the woods behind me.

I squinted through watery eyes, trying to distinguish
the line between field and sky. It wavered beyond the
corn, undulating in the heat. I tried to remember if
any of the ghost stories had song attached to them like
Harriet Tubman. The sweet chariot wouldn't have
to swing too low on the Eastern Shore where the sky
seemed to swallow the puny earth.

My mother taught my family American folksongs
that she had learned at Episcopal youth camp. I turned
off the radio and sang *I've Been Working on the Railroad* to
the Eastern Shore farmhouses, just to see if I could still
get through the whole thing.

> *I've been working on the railroad all the*
> *live long day,*
> *I've been working on the railroad just to*
> *pass the time of day.*
> *Can't you hear the whistle blowing?*
> *Rise up so early in the morn!*
> *Can't you hear the captain shouting?*
> *Dinah, blow your horn!*

You can hear the swing of the hammer in that song.
The swinging singing made me feel better, and the song
tied me into some history somehow.

Carl Jung was right; exploring folklore and its
archetypes can make us happier humans, more
understanding of our temporal plight and more

appreciative of our community of stranded mammals, cognizant of our own deaths. Story holds a people together, like the Native American Indians clutching to xenophobia to maintain their diminishing culture.

Each generation retells the same myth, but in their version. We keep buying into the myth over and over again, with the simplicity of children, like someone hypnotized by a flashing light. Mythology's hard wired into our heads and through our hearts, tugging our heartstrings and tear ducts, and we don't even see it coming. Our group reaction to the stories is comforting. We make patterns. We make archetype.

Could archetype ameliorate our ultimate struggle with our temporality? It's going to happen. We all will die. As humans, we are all terminally ill. Psychiatrist Elizabeth Kubler-Ross pioneered the counseling of terminally ill patients, helping people deal with the inevitable fact of their deaths. She qualified five stages of terminal confrontation: denial, anger, bargaining, depression and acceptance. Maybe this American civilization is still stuck at the denial stage. Certainly the town of Cambridge is fixated at that step.

Once we as a culture lose the terror attached to our mortality and begin to see the bright core of our spirituality, our myth will change. Maybe as the baby boomers age and a greater percentage of the population grow over fifty, America will mature past its current adolescent stage of war, violence, youth obsession and short-term attention span. If only we could think beyond our linear concept of time. That's the true curse of temporality: being so linear. Maybe one fine day we'll understand that there is a magical energy inside all of us that does not end.

And maybe the beginning of that understanding is believing the miracle of ghost stories all around us.

Feed your brain on archetype.

Driving through the Eastern Shore percolated and simplified my thought.

I shot through the woods in southern Talbot County and sensed a weird lucidity to the world. The dimensions seemed sharper and the colors brighter. I was dizzy with clarity. I rolled down the window and called out, "Turn off your TV and go out into America. Use the good side of xenophobia!"

According to a June 2003 Harris Interactive poll, fifty-one percent of Americans didn't think they would take a summer vacation. In these tight financial times, we should all vacation within a 50-mile radius and revel in our state-ism.

"This is my call to localization: drive and explore!" I called out.

The grizzled, old man in a green pickup next to me gave me a funny look and turned up his stereo. As he adjusted his fishing hat and sped up to pass me, I thought I could hear him think: *she ain't from around here.*

Learn your local history; it's close by. Be an amateur

scholar, I thought.

Volunteer at the library.

Believe in pirates.

Find out which statue generals have one leg up and why.

Walk a ghost tour.

Ask your friends about their ghost stories.

You'll be amazed what phantoms you'll find in your own backyard. In North America, all land can possibly be Indian burial ground.

We're only allowing ourselves to see the middle of our vision. Put the blinders down and look at what's on the sides. It's scary but it's what's really there.

Hear the folklore from your neck of the woods.

As I drove out of the Talbot woods and the Choptank River curled before me, the clouds that had cleared at the Bay Bridge resurfaced, building along the river and rumbling and shifting over the top of the trees. Dorchester County was holding its breath for another storm.

The car bumped onto the Malkus Bridge. The tires clacked over the supports. I counted the beams over its 1.5 mile span.

We like to count, I thought.

People have been counting the Great Ages since before the birth of Christ. The Great Ages are not geological ages but plotted on the movement of the

Great Constellations, a measurement of stellar reference points. Currently, by a variety of estimates, we are transitioning from the Age of Pieces into the Age of Aquarius, from the age of patriarchal protection to independent thought. This current time is the conversion between the paradigms, and we're both lucky and cursed to live through it. We're lucky to possibly contribute to the shift but cursed because we probably won't live to see the outcome. The Age of Pieces theoretically began in 60 B.C. The beginning of the Age of Aquarius ranges from 1967 to 2100 A.D. and will usher in the decline of religious power and the increase of individual revolution. It marks the end of times and the dawn of a new world order. Scientific sacred cows will fall. People will think for themselves. Something magical will happen with water and air.

If you believe that stuff, I thought, as I bumped into Dorchester County.

Today

BURIAL RECORDS

As I parked the car by the courthouse, I thought of William Shakespeare's line from his play *Macbeth:* "so foul and fair a day I have not seen."[86] The weather was weird and wild; a waltz of both stormy and sunny. Towards the middle of the sky, pasty clouds began to block out the blue. Towards the northern horizon, sallow clouds transitioned to gray and then greenish black, looking almost yellow along the horizon. The wind was up, and the barometric pressure was down. My knees throbbed. The air was charged with something that felt like electricity. I could smell the brine and fish from the marina. Seagulls flapped above me. The Spring Valley grass sighed under my boots. The fountain splashed. A songbird trilled in the graveyard.

The Dorchester County Courthouse sits back on its spongy plot; its two light-coffee, brick levels resting on a row of Ionic columns. The wind whipped the American and Maryland flags at the top of the courthouse steps, and the lanyards on the marina sailboats clanked. Three locked, brass doors faced High Street, and the lush lawn beside it sloped down to the creek's edge. Down the hill, at the end of Court Street, a much more rectangular brick building housed more county offices. Its parking lot was edged on two sides by the retaining wall of the creek. The cove water level looked low, and I wondered if the tide was out. Dorchester County uses the side entrance of its courthouse as its main egress. Next to the metal side entrance door was a sign that said: No cell phones and no weapons. I shouldered and pushed hard on the door and found myself in a short, sky blue hall with a metal detector and uniformed guard. The building's interior was cool, and the place had the feel of being freshly renovated. My boots scuffed at the stone-gray, low pile carpet, and I almost tripped into the metal detector. The droopy man in a beige and brown uniform shifted his weight on a wooden stool. His moustache was laced with breadcrumbs. Pipes clanked distantly in the bowels of the building.

"Construction records?" He parroted in response to my request, raining crumbs. He checked his clipboard. "That would be in the Register of Wills, I suppose. Down the hall and to the elevator. Down a floor," he said. He seemed tired.

I tiptoed down the carpeted hall to the elevator. A legal assistant scuttled by with a stack of thick, manila folders. The sound of something printing clacked from an office down the hall.

"Looks like it might rain this afternoon," the younger man said to the guard.

"Again. It'll flood the basement again," the guard said and then muttered something about wet boxes.

The elevator was modern and shiny and only went down one floor. It opened to a basement hall with three door options: two were marked as No Admittance and the third was the Register of Wills. I realized that I was in the end of the building that the courthouse book claimed had been renovated in 1931. I opened the door to the Register of Wills and a bell rang, like at the Dorchester Arts Center. I faced a long, mahogany counter. I wondered why, after the 1852 fire, the will records weren't safely enclosed in a vault riddled with fire sprinklers. The Register had more robin egg blue walls, several desks and some potted plants. Back towards the creek hulked stacks of filing cabinets under square windows. Each desk had a visitor's chair, and an old man in a black suit and a bowler sat in one back towards the filing cabinets. He seemed to be waiting. A cotton-haired woman in a floral print dress squinted out at me from behind the counter and a chunky pair of reading glasses. I wondered if she was Dorothy who never called me back. Her arthritic hands trembled slightly, and her pale eyes were coated with cataracts. I wondered if she was a cursed LeCompte.

She spoke to me through her nose. "Jail construction?" She responded to my request. "Well, you could look in the Maryland Room of the library . . ."

"I've been there. There wasn't much, well, one book on the courthouse, but it doesn't fit with the other stuff I've been told."

"Well, that would be history," she said. "I guess you could look in our history files. I doubt that there's anything that old in them. We've had several fires, you know." She slowly directed me to a wall of sagging metal filing cabinets. The smell of old books hit me in a wave. The air was syrupy with story; it required a greater effort to breathe. "I never heard of an Indian burial. I thought you meant the renovations." The old man gave me a wincing look, but the cotton-headed Register lady ignored him. "The 19th century's in that drawer," she said. "It's not much."

I dug through the 19th century files for a half an hour. Sometimes it was hard to focus because my eyes were so watery. Through high, square windows I saw the greenly darkened sky. My right knee felt stretched with water and popped.

It'll storm again soon, I thought.

I found nothing about jail construction, but I did discover that Dr. Elmer R. Reynolds, who donated the human skull to the Smithsonian, was a resident of High Street in the later part of the 19th century. He had lived on the creek side of High Street; he could've found that human head in his backyard.

I was replacing the files when I saw a piece of yellowed newspaper wedged in the back of the file drawer. It reminded me of the Willimina research page I found stuck between my pasta and towels. I carefully extracted the paper. It scraped against the metal of the filing cabinet.

Could the clerk hear it? I thought. It seemed so loud.

It was an article, torn from some newspaper. It didn't have a date or masthead; the top had been ripped away. *Freak Accident at Courthouse*, the title read.

The article described a cave-in during the excavation of the jail basement. No one was really sure what happened and in what order, but witnesses said the earth collapsed, leaving a wide gap that let in Cambridge Creek. Very quickly the creek flooded the basement hole with waist-deep, swirling tobacco-colored water. One witness, a Dr. Elmer Reynolds, thought that the water developed a current. The workers scrambled to reach the wooden ladders, but the flood was too powerful. The water suctioned men out to sea, out through the hole and through a roaring tunnel. Their bodies were spat into the creek.

Just like in the Christ Church slave story, people die in basements here.

The article listed the missing men. They were all Civil War veterans; the jail construction took place during Reconstruction. Everybody needed work. One of those workers dragged out to sea was John Austin.

How many John Austins could there be in one town at one time? I thought. It must be Big Liz's murdering, plantation boss.

Towards the end of the article, the writer reported that several boxes of "fantastic fish bones" were saved for delivery to the Army Medical Museum. Was The What a big fish? Did Reynolds get an Army promotion for delivering an Indian head, I wondered.

I stared at the newspaper. A corner of it flaked off in my hand. I was afraid I might breathe it in. These were the men who disturbed The What; these were the pot-diggers who opened up Pandora's Box.

I'd been trapped in a rip tide once as a child. I imagined the creek current tugging at me, pulling me to the hole gaping in the basement wall, my bloodied hands clawing the mud, desperately grabbing for anything to hold, hearing screams as I'm dragged under and sucked into the torrent of the tunnel, no air, water crushing my bones, shot out to sea, no air, swallowing water, breathing water, the hum of bubbles in my ear.

How could I put those watery deaths into the tour? I thought.

"Check the burial records," rang Maynard's advice in my head. "Scant records remain."

Check the burial records, not the construction records. At the end of the article was a list of workers who had survived the incident. The courthouse building seemed to dip a little towards the creek behind it. The high windows sank.

"Did you find something?" The clerk called out to me.

"No," I lied. I tucked the newspaper into my papers.

"That's a shame," she said and walked into a back room. "Not much there."

I heard her moving boxes and opening drawers. In her absence, I copied the survivors' names and stuck the newspaper article back into its hiding place. I gathered my books. Thunder rolled across the sky. The cabinets vibrated. The old man coughed.

Was no one helping him? I wondered. What was he waiting for?

"Looks like it might storm," the clerk said, returning to her post.

My chest tightened. I felt light-headed.

"I'm looking for the death certificates of each of these men." I handed her my list. "I know they died around Reconstruction."

She read the list. "This might take a while. Have a seat. You got here just in time. We close at four thirty."

I sat in a visitor's chair and wrote a quick poem while I waited for the clerk to pull the death records. I was wrong. The courthouse had part of the answer to the mystery, and ghost stories were the other part. I

felt so close to an answer, but none of us will get close
enough to the mystery to understand it completely.
We're too linear, crippled by our literal brains, applying
anthropomorphic reasons to the non-corporeal.

The clerk returned quietly. "Thanks for your
patience. I'm sorry I took so long." She smelled of
vanilla cookies and sweat.

I rose and looked at the clock. Almost an hour had
passed. How? I thought. It only felt like a few minutes.
The old man was gone. We were alone. Where did he
go?

"I only found John Austin and Elmer Reynolds."
Her hands were shaking. "The others didn't even have
birth certificates. I don't know where they came from,"
she said. "Look at this." She held out her notes. The
page trembled. "Both of them died by water."

Reynolds and Austin survived the cave-in but later
died from deaths involving water. Reynolds drowned
and some said that Austin was lost at sea.

"There's a lot of water in this town," she said.

There was a puddle on the chair where the old man
had sat.

"Thank you," I said and aimlessly took the list. The
old man had disappeared, leaving only water behind.

A flash of lightening etched across the darkening sky.
"That one was close," she said, shaking her head. "They
said severe thunderstorms this afternoon."

"Thank you," I said again. The old man must be a

ghost, I thought, and made of water.

Water was the enemy, and it covers most of the
planet. Even though we're made of it, it can drown us.
Water killed the pot-diggers. Water featured in much
folklore: in Hurricane Hazel, in Big Liz's swampy death,
in Wallace's swampy death, in ghost ships and pirate
tales, in the sacred Indian burial site. Water branded my
cheek, stained my wall, and tore at my house foundation.
Water flooded the Cambridge basements and gurgled
in their walls and steeped in the grass. Water was
swallowing up inches of shoreline nightly.   Water could
be the reason for all that Cambridge denial.

I felt an odd hum underneath a crack of thunder,
like the din of a multitude of voices far away. My vision
darkened and brightened.

"Are you all right?" I heard the clerk ask.

I felt something liquid dripping down my cheeks
and onto the counter, bouncing off the glass top. The
floor vibrated slightly as if a train was heading towards
the courthouse. I wiped the tears away. "I thank you,"
I said a third time, like a spell. "The power of water,"
I muttered as I staggered out of the Registry of Wills.
"Where did the old man go? Was he Elmer?" I couldn't
wait for the answer or the elevator and lurched up the
stairs.

I could hear the woman calling after me, up the
short flight of stairs, "Miss? You dropped a paper!"

I stumbled up the polished stairs, down the indigo

halls and past the ancient guard.

"Did you find everything you wanted, miss?" The guard asked.

I heard someone say to the guard, "Marvin, look at that sky."

The guard replied, "I hear we're in a tornado warning. Maybe we should get out early and get home."

I pushed out of the courthouse door and leaned against the exterior wall, panting and sweating as if I had run a mile. The air was heavy with humid electricity; a storm was coming. I smelled ozone. Lightning pulsed in bulky, green clouds towards the river. I found paper clutched in my hand. It was the poem I had written while I waited, in the time that felt like a minute but was really an hour. I wondered if the old man was a ghost because the clerk never acknowledged him. He had weight like all the other old man ghosts. A Styrofoam cup flew by. The sign by the creek that read No Fishing No Crabbing swung and clanked. A gust of wind turned all the leaves on the courthouse trees backwards and pulled at my dress. The sky was emerald and purple. I sat on the garden wall by the courthouse and pulled out the Great Choptank Parish book that Iris had given me months ago.

Something about the graveyard, I thought.

The air was so sultry and hot that it was hard to take a full breath. Something hit my face as it flew by. It seemed white. The wind blew the pages of the book to

the graveyard section in the back. In a photograph of
the graveyard from the 1930s, clouds were building up
behind the trees, as if a storm was coming. A woman
walked out from behind a towering gravestone, dressed
in Victorian garb, carrying a baby. She got closer,
walking slowly towards me; her face was mine when I
was twenty-eight.

How could a figure move in a photograph? Was I
related to Willimina?

I blinked, and like the serpent in the clouds, the
images shifted. I blinked and almost in stop action the
woman seemed closer and suddenly had a dark stain on
the front of her dress. I blinked and behind Willimina
and her dead baby leered the old man from the Register
of Wills, tipping his wet bowler.

I shrieked, dropped the book and ran down the hill
to the water's edge.   I felt pulled to the creek. The water
in the cove was rapidly draining, like the water out of
a tub. The boats were sinking to the creek floor. Still
tied to their slips, they began to either pull down the
moorings or hang alongside them. A man in shorts and
a white tee shirt tried to climb up the slippery pylons as
his boat sank to the bottom.

This is not right, I thought. This is how harbors
behave right before a tsunami.

The ground began to shake, like a train was coming.
The wind howled a weird suction sound. Tiny pellets of
sharp hail spattered my hair and stung my cheeks.

Near the horizon, down past the condos and the seafood company and the restaurant, down where the creek met the Choptank River, murky clouds were spinning, spinning down to the water in a furious funnel, rotating counter clockwise, drilling into the creek. Something in the creek reached up to the funnel, and the water level dropped, down and up in a vicious updraft, up the wavering funnel and out to sea. The air temperature dropped ten degrees. The wind blew in harsh gusts at me but my leg muscles locked. The force tore at my clothes. It ripped my notes out of my hand. Dirt whipped through my hair. The hailstones grew larger. The darkness of wind over water painted the remaining creek black.

Get under cover, I thought.

The suction noise shifted into the scream of squealing pigs, roaring lions and jet engines all converged into one horrible wail. A downpour wall of water advanced from the Choptank behind the vortex, a sheet of water as high as a cloud, heading towards the funnel and me. I was riveted by the staggering power of that cone cloud, a witches' brew that churned up the sky, tossing lightening down to the river's writhing surface. Somewhere in the algebra of its cone was the Divine Proportion of PHI. The frothing, swirling whirlpool at the river's mouth rose up, slowly, gracefully, gathering more bay and power with it. As the sheet of water passed over it, the waterspout balletically rose up to the river's surface and hovered and waited. The wind shrieked, and the funnel raced towards the courthouse, barreling like a train, a whirling tornado of water and boardwalk and boats. It picked up a storage building of the seafood company; picked it up like a toy and tossed it into the High Street backyards. I was rooted. I couldn't move.

This waterspout's not carrying frogs, I thought.

"I've seen a water spout," Debbie had said. "They're serious."

Run, I thought. Get out. Get out now.

I heard people shouting behind me as pieces of lumber, shingles, insulation and debris flew by my head. The noise was deafening. The wind uprooted trees and blew bushes down High Street. Something hit my shoulder.

Cambridge Creek had disappeared, and the fish were stranded on its sticky floor, flapping and writhing. They looked like mermaids with sinewy arms and finned tails below their waists. They seemed to be moving in a pattern of sorts, trying to walk in a circle around a dark muddied form on the cove bottom. When the last bit of water and sludge and slush drained away from it, I saw bones, bones in a perfect shape of a massive sea serpent, twenty-five feet long and stretched out and resting on the marina floor. The strong smell of fish wafted up.

Fantastic fish bones, I thought.

The bones began to move. The skull moved first.

Then out flipped a foot. Or was it a hand? A pair of wings undulated on its back.

I screamed. The tumblers of the universe clicked.

The waterspout hit me. My mouth was open and filled with bay. A bright light flashed. For a dizzy moment, the tornado spun the world around me, a wiggling curtain of river packed with houses, buggies, headless slaves and coffins. Like my dreams, I saw Hannah in her shroud and the old sea captain fly by. The lashing sheet of water passed, and for one shining flash I balanced in the center of the storm, a cone of whirling sea surrounding me. I looked up and saw an undulating circle of perfect blue sky above me, a wormhole to another time. Lighting laced the funnel, and the wind picked me up and felled me to my knees. I clawed at the grass clumps at the creek's edge. I clutched the soaked earth. I felt weirdly clean.

When the curtain of waterspout passed, the air was clear and the sun was shining on the other side. It happened so fast. I stood up slowly. My knees felt watery but had stopped throbbing. I rubbed my moist eyes. I could see clearly.

The sea dragon serpent bones on the creek floor raised its head. Its flesh was restored and it was intact. Its scales glittered in the sun. Its muscles rippled under its scales. Two ribbed wings flexed on its curved back.

I thought, those fish are not fish.

The Japanese believe that dragons, gods of water,

live under the sea in a world without time. The Chinese believe, like the Native Americans, that dragons bring rain and escort people to heaven, that a dragon's breath is *sheng chi* or the essence of life.

I stopped, noticing something on the riverbank even stranger than the sea dragon and its mermaids. At the end of its wriggling tail was a dark revolving door set into the earthen bank. It looked like the entrance to a prestigious bank. Its frame was burnished metal, gleaming in the sun. Its glass was polished black. Roots grew around it, woven through its metal casing. Its glass doors vibrated slightly. It seemed to hum.

Something was wrong.

The sea dragon opened its mouth and moved its lips. I heard what sounded like Maynard's voice in my head, his well-modulated, calm voice that was clearly not my own.

I thought I should be more freaked out about a voice in my head that was not my own, but it was so soothing.

The sea dragon said, "Do not fear. Tell everyone to learn the stories so they are prepared. Spread the word. Everybody dies. Dying is only a transition, little foreigner."

This is not real, I thought. You're sick or delusional or the waterspout knocked you out. You're reflecting on this experience in your own images and archetypes. That way you can handle it. Like dreams, like parables, like ghost stories, like the sea monster sketch in Flowers' book, like the Indians offering fish sacrifices, like the archetype of the Water Spirit.

"You are not sick or dreaming. Tell them!" The sea dragon implored, shaking his mucky locks towards High Street. He shook his ears out of his head and they flopped and perked up. His beard shuddered out of his chin. The mermaids danced on point around the sea monster. "Look, little foreigner!" He cried in my head. "Look!"

I turned and saw Jane, conducting a ghost tour to a small group of six tourists, one older couple and one younger couple with two kids. Jane and her tourists seemed incredibly dry, even hot. Had they been inside?

"Did you see that storm?" I tried to say but my mouth wouldn't move. It was too wet.

Fear is nothing, I thought. Yet it gripped me in its cold arms.

Jane stopped her group besides me at the water's edge. She ignored me. "Let's stop right here," she instructed her group. "Everybody gather around," she said softly.

This spot isn't a stop on the tour, I thought. She's supposed to stop further up by the sidewalk at Spring Valley.

I noticed several downed trees behind her and the courthouse seemed to suddenly be under construction. Part of it was missing and scaffolding, like the photos from the 1970 bombing, ringed one corner.

"At this very spot, the Baltimore writer who wrote the ghost tour mysteriously vanished during a freak storm and was swept out to the Bay. They found her body all the way down by Elliott Island in southern Dorchester County. All of her clothes were removed and she was coated with snake scales," whispered Jane. "Months later they found her boots in St. Mary's County across the bay."

An old man in the crowd looked dubious and turned out to look towards the river. He adjusted his baseball cap. He looked a little like the old man in the Register of Wills in different clothes.

"I read about that in the paper," said the younger woman, fanning herself with a brochure.

A small boy's mouth gaped and whistled. "Nasty," he said. He nudged the old man "Did you hear that, Gramps? That's the best so far."

"This isn't funny, Jane," I said darkly, but no words came out.

Jane didn't answer. A hot summer evening breeze ruffled her hair, and she wiped away a tear. "She was a nice lady. People say the water spirit got her."

I cried, "I didn't write that!" But the wind in my ears swallowed the sound. "Jane! The burial site's not under the jail. It's the dragon in the creek! Look in the creek!"

Jane walked away from the group, towards the graveyard, obviously grieving. The tourists mumbled and exchanged tense looks. They clutched their gift shop bags and followed her.

"What's wrong?" The kid asked.

"Shh!" The old man hissed.

The younger man photographed the marina. No one mentioned the twenty-five foot long talking sea dragon or the dangling boats or the door in the creek wall.

As Jane walked past the graveyard gate, Harriet Tubman glided out and held out her right arm to me. I thought she was an actor from the Tubman Museum or maybe that Judy had contracted ghost actors until a pillar of fire sprung out of the ground beside her. The column glowed and contracted and pulsed, a ten-foot cylinder of green and gold light seemingly growing out of a grave behind her.

Harriet's come to help me cross, I thought. No, I'm dreaming. No, I'm not.

Someone was in the room in the half hour before my grandmother and great-grandmother passed. In the half hour before both their deaths, the sky in their bedrooms became thicker and cloying with so much more energy that sparks lit the air. That's why the air felt so charged when I went into the courthouse. It wasn't the storm. It was Harriet.

Einstein was right. Space and time are bendable and light is the constant.

I heard the doctor's voice call, "Show me the way!

Show me the way!" I heard the buggy wheels creaking and the horses' hooves pound.

The black shapes I had been seeing in my periphery zipped around the graveyard walls and spun like mini tornadoes around the graves, turning golden as they spun. The yard was filling with ghosts and the warm amber light of the Other Side. Big Liz walked out from behind a tree, holding her head in her broad arms. She was wearing a big tobacco knife. The lips on her decapitated head pursed and blew.

Oh my God, I thought, the stories all are true. They really happened once. The stories are saying goodbye. They're staying behind, to teach, to tell.

"The Veil is ready for you now, little one," said the sea dragon.

The hard man from the teddy bear store in Fells Point stood next to Hannah Maynider, wrapped in a tattered shroud. Beside them stood a young woman, holding a bloody baby, a young woman who looked like me. Willimina. Horn's Point. How I knew the back roads. Why I was here.

"You can only completely understand if you cross the Veil," Maynard had said.

No, I thought, I want to live to tell the tale. Someone wake me up!

Cannon booms rang over the empty creek and up and down the Choptank. Someone was dredging the river.

In the graveyard, someone began
singing
"Maryland, my Maryland."
"Avenge the patriotic gore
That flecked the streets of Baltimore,
And be the battle queen of yore,
Maryland, my Maryland!"[87]

Be the battle queen of yore, I thought.

Judy had called my name across the graveyard in the same way in May. The spirits' bodies turned glossy and wavering and then began to sink into the ground.

My feet turned to water. It was a sudden transformation and felt rather like a melting, a spreading out of my toe atoms into the ground, like the casual suspension of a thicker liquid into a thinner one, like cream into coffee. The dissolving was not like falling into a marsh hole or being pinned to the earth in the Atlantis dream. It was graceful and wrapped in comfort. I sunk into the grass and slowly drained into the earthen bank: my legs, my knees, my hips, my pelvis.

I opened my mouth to scream because the sensation felt so strangely gurgled and bubbly and almost ticklish.

Will I miss my feet, my legs, I thought wildly, and as quickly as that thought I was waist high in earth and water.

As a fetus, my lungs were once filled with amniotic fluid and for a fleeting moment I could almost remember that feeling. I remembered the rip tide claustrophobic sensation of drowning and my lungs filling up with water. But my lungs were not filling up with water, they were becoming water, and the air pushed out so fast that I didn't miss it. The dissolving felt weirdly good, a welcome release, like removing a pair of tight shoes after a long day. I felt my muscles separate. My hair spread out into the grass as I percolated into the soil like the tomato water. I felt the roots of the trees and the wriggling of insects and worms. I knew the maze of dirt. I was pulled to the marina like metal shavings to a magnet. I no longer felt pain. My knees no longer hurt. All that leaking and shower absorbing and night sweats had prepared me. I was living my Atlantis dream. I was water.

The revolving door began to spin wildly, sending off sparks, and the marina filled back up with creek and me.

I am the top of a puddle. Sometimes I am a raindrop on a map. I am standing water. I am the wave around the dock. I am rattling pipes. I am a bubble of beer moving through a man's gullet. I am light on skin, turning melanin tan. I am sleet bouncing on the tops of cars. I am thunder rolling over a hill. I am lightning kissing a tree.

I love the freedom; it is pure beauty, more beauty than words can describe.

I hear lectures on historic inevitability, the metaphysics of weather, and the movement of peoples across landmasses. The collective ruminations are not too far from my long ago individual thought, but the lecturer is either the sea dragon or Maynard.

Why you? I wonder of the monster.

"Because you created me from archetype. Because you expected a symbol like me. Because we live under the seas in Atlantis. Because Cambridge is a Thin Place. Fog happens when warm moist air rolls over cold sea or when warm land loses its heat to cooler air," he calmly drones on.

I am breeze. I am joining the Gulf Stream. I am a tendril of fog around a tree branch.

"The Civil War will be unimportant in 500 years. Nation states won't matter. State-ism and xenophobia will wane with cross-pollination. Migration will mix accents and blur geographic boundaries. Mankind is melting the ice caps and making more room for us."

I am a moonlight beam on a tent wall. I am the refraction of a rainbow. I am the bubble in a waterfall. I am the water molecule holding up a bug's leg as it skates over a pond surface.

"Your soul leached from the soil into the water table and into the river where it can move freely. Your consciousness and your soul are now weather. They are water spirit. This universe is a closed system. Energy does not end. Do you understand, little one?"

I try to answer but my lips are in the bottom of the ocean. I oddly don't care.

When I move through the wharf pilings and whip around boat bottoms, I am linked to other weather molecules, other souls leached from the graveyard's bounty. We are a necklace of weather. We are the dance of a storm front. We weave a filigree of fog over Dorchester County. We are downpour. We are turning the planet on its axis.

"You are all together, all the ghosts from the stories," says the sea dragon. "History equals myth."

I am bouncing balls of hail. I am a mote of dust in sunlight. I am a teardrop, rolling down a scarred cheek.

"Living humans share a communal collective of myth in order to ease the eventual death transition to energy group consciousness. Ghost stories are told in archetype so man can become accustomed to the eventual shared consciousness."

I swim around a rock in the river. I dodge a seagull. I am the water in a bucket of tomato plants, sitting on a back porch. I am a water stain on old lead paint.

"Religion, music, art, theatre, sports prepare mankind for the eventual group consciousness. That's what catharsis is – a fleeting moment of shared consciousness."

I wonder if I'm dreaming or died or if I am in a summer coma or if I fell into a story, but as time swirls and spins on itself like weather my own thoughts spin off with it, twirling into the Divine Proportion.

I am no longer waiting for someone to say, *Maryland,*

hon, wake up.

"Wind is a transference of ions from positive to negative and back again, constantly," the sea dragon says. "Constantly, do you understand? Stories move that wind. Faith and prayer and hope move that wind."

I am mist over High Street when suddenly it all makes sense. I remember Hannah's grave robber story and when I remember a story very hard, the wind blows through the trees and the current flows through the channel.

I am water streaming through a dredger's net. I am the splash of fish thrown into the creek.

"You're new but learning quickly," the voice congratulates me. "You have nothing to do but learn and nothing but time to do it in."

I want paper. It's the first want I've had since the waterspout. I try to remember where I last saw stacks of paper. Where is the closest place with stacks of paper?

I found myself in the upstairs hall of the Dorchester Arts Center outside Delia's office.

It is night. I hear someone trying to sneak down the stairs in the half-light. I push myself up against the hall wall. A tall woman in a jean skirt and a black sweater walks unsteadily on to the landing. Her knees are watery. She's obviously frightened. I can smell her fear. Her fear makes me stronger. I hold the wall.

She looks familiar. She looks like Willimina would if she had reached forty. She wears glasses and a purse and blinks at me. She leans into the wall to investigate the stain I make on the paint. She reaches a tentative finger up to touch and I remember the old man in Hurricane Hazel and water blows into her face. Some lands on her cheek, and she rubs it away. She looks wild-eyed and turns, clinging to the railing and reaching towards the stairs to the first floor.

"Maryland?" A reedy, older woman's slightly Southern voice floats up the stairs. "Maryland? Is that you?"

It's me. The tall woman in the jean skirt is me. From the water stain on the wall in late May, I watch myself stagger down the stairs. The sanders whine in the back of the house.

"You can do what you want but the outcome's gonna be the same," she mutters in my old voice.

Maynard had said, "If you have crossed The Veil and are centered, then the beginning has no end."

The beginning has no end. I wish I could cry. I can't call out. I can't chase her.

I have to tell somebody. There's no need to fear.

I pull myself off the wall and remember the paper on Delia's desk. I'm in her office. I try but I can't pick up a pen. I can't pick up the paper. When I push against the paper, the parchment turns to wet mush. Delia's in-box is full of water. I try to write and I soak the page. I'm exhausted and wait.

I hear the voice. "The thought of writing is

enough." The night passes through me. "Waiting is easier now for time is not linear. You're beginning to understand."

The wet paper beneath me is almost a downy mattress.

"The stories live on for the living. Myth's more important to the living."

In morning, Delia clumps into her office and thuds her coffee cup on the desktop. The motion wiggles me. She leans over and peers at me, resting in her in-box, soaking all her files.

"What the hell?" She mutters. "Judy! Why's my in-box soaked?" No one answers. Delia carefully balances the box and carries it to the second floor bathroom. She pours the excess water down the sink drain, and out I go to sea. I clank the pipes as I exit the building. It's a fond farewell.

I hear Maynard's voice say, "When you've been here longer, you'll understand."[88]

I can't remember when I heard that line before but it rings in my mind like an archtypal knell.

After April 2nd
WE BELIEVE

Jane stands on the bank of Cambridge Creek, the wind blowing her straight, brown hair like a flag behind her. A gray sky whirs overhead. Low clouds scuttle over the tops of the boat masts. Twelve tourists in bright raincoats stand, watching the boats clink and thud against the moorings. A teenage boy stands beside Jane. He is a holding an iPod. He is learning the tour.

"And when Delia got to her desk the next morning, the morning after the storm, her in-box was full of water," whispers Jane. "And one of those papers was the original copy of this tour."

A lone raindrop hits a ten-year-old girl on the end of her nose. Her mouth slacks open with wonder at the stories.

The teenage boy reads from his iPod. "A couple of days later, the Courthouse guard found some of Miss Maryland's notes on a crumpled piece of notebook paper. Her relatives confirmed her handwriting. The paper says this. 'If you see something you can't explain, accept it and try not to repress it. If you hear something scratching down the sidewalk on a windy night, listen. Don't be surprised that there are many sides of the world we call reality. Listen to the wind and watch the leaves for I am in them. I am the top of a puddle. I am a raindrop on a map. I am lightning kissing a tree. There is nothing to fear."

Some tourists look pleased, others vaguely confused.

"Her brother found this diary in her effects. Some people doubt its authenticity because the last entries are not possible," slowly says the boy, lingering on each word.

"But we believe," says Jane seriously. "Because the last entries were damp."

The ten-year-old girl slips her hand into her mother's. Her mother rolls her eyes to the scuttling sky.

"That's more alive than a monument with two legs rearing up," reads the boy quietly to the lowering heavens. A breeze wiggles through his windbreaker. "I've become story. I've become myth."

"I am thunder careening over the Bay. I am wind. I am rain. I am." Jane smiles sadly and closes her book.

Endnotes

[1] Website for the Dorchester Arts Center

[2] James Ryder Randall, *Maryland, My Maryland*

[3] John R. Wennersten, *The Oyster Wars of the Chesapeake Bay*

[4] *The Baltimore Sun*, December 1945

[5] National Bohemian's beer campaign for Maryland

[6] Morris L. Radoff, *The County Courthouses and Records of Maryland, Part One: The Courthouses*

[7] Elias Jones, *The New Revised History of Dorchester County*

[8] Elias Jones, *The New and Revised History of Dorchester County*

[9] Thomas Flowers, *Shore Folklore* and *A Young Person's History of Dorchester County;* George Carey, *Maryland Foklore* and *A Faraway Time and Place, Lore of the Eastern Shore;* Vernon Griffin, *The Veil and More Folklore of the Eastern Shore,* Brice Stump, *Stories of Southern Dorchester County*

[10] Thomas Flowers, *Shore Folklore, Growing up with Ghosts, 'N Legends, 'N Tales, 'N Home Remedies*

[11] Brice Stump, *Stories of Southern Dorchester County*

[12] Thomas Flowers, *Shore Folklore*

[13] Brice Stump, *Stories of Southern Dorchester County*

[14] Thomas Flowers, *Shore Folklore*

[15] Webster's Dictionary, Second College Edition, 1974

[16] Oxford Dictionary, 1998

[17] Webster's Dictionary, Second College Edition, 1974

[18] George Carey, *Maryland Folklore*

[19] Thomas Flowers, *Shore Folklore*

[20] Webster's New World Dictionary, Second College Edition, 1974

[21] George Carey, *A Faraway Time and Place*

[22] Donald Shomette, *Pirates on the Chesapeake*

[23] Vernon O. Griffin, *The Veil and More Folklore of the Eastern Shore*

[24] John R. Wennersten, *The Oyster Wars of the Chesapeake Bay*

[25] *Intimations of Immortality*, William Wordsworth

[26] Joseph Campbell with Bill Moyers, *The Power of Myth*

[27] Joseph Campbell with Bill Moyers, *The Power of Myth*

[28] Thomas Flowers, *Shore Folklore* and Dickson Preston, *Trappe, the Story of an Old Fashioned Town*

[29] Thomas Flowers, *Shore Folklore*

[30] John S. Hill, edited by Harold Roth, *Stories of the Eastern Shore*

[31] Thomas Flowers, *A Young Person's History of Dorchester County;* Helen Chappell, *The Chesapeake Book of the Dead*; George Carey, *Maryland Folk Legends and Folk Songs*

[32] Thomas Flowers, *Shore Folklore*

[33] Thomas Flowers, *Shore Folklore*

[34] Joan Didion, *The White Album*

[35] Thomas Flowers, *Shore Folklore*; and Helen Chappell, *The Chesapeake Book of the Dead*

[36] P.L. Travers, *Mary Poppins*

[37] Lillian Jackson Braun, *Short and Tall Tales*

[38] Dorothy Parker

[39] Joseph Campbell with Bill Moyers, *The Power of Myth*

[40] Sarah McLachlan, *Drawn to the Rhythm*

[41] Trish Gallagher, *Ghosts and Haunted Houses of Maryland*

[42] Trish Gallagher, *Ghosts and Haunted Houses of Maryland*

[43] Trish Gallagher, *Ghosts and Haunted Houses of Maryland*

[44] Brice Stump, *Unforgettable Treasures, People, Places and the Culture of the Eastern Shore*

[45] Brice Stump, *Unforgettable Treasures, People, Places and the Culture of the Eastern Shore*

[46] William Shakespeare, *Hamlet*

[47] Thomas Flowers, *Shore Folklore*

[48] Gary Zukav, *The Dancing Wu Li Masters, An Overview of the New Physics*

[49] *The Dorchester Star*

[50] Thomas Flowers, *Shore Folklore*

[51] Vernon O. Griffin, *The Veil and More Folklore of the Eastern Shore*

[52] George Carey, *A Far Away Time and Place*

[53] Colonel James Sulivane's 1925 High Street article

[54] Helen C. Roundtree and Thomas E. Davidson, *Eastern Shore Indians of Virginia and Maryland*

[55] Helen C. Roundtree and Thomas E. Davidson, *Eastern Shore Indians of Virginia and Maryland*

[56] Helen C. Roundtree and Thomas E. Davidson, *Eastern Shore Indians of Virginia and Maryland*

[57] Elias Jones, *The New Revised History of Dorchester County*

[58] Thomas Flowers, *A Young Person's History of Dorchester County and Shore Folklore*

[59] Elias Jones, *New Revised History of Dorchester County*

[60] Arlene Hirschfelder and Paulette Molin, *Encyclopedia of Native American Religions*

[61] Edith Hamilton, *Mythology*

[62] Thomas Flowers, *A Young Person's History of Dorchester County*

[63] Arlene Hirschfelder and Paulette Molin, *Encyclopedia of Native American Religions*

[64] Joseph Campbell with Bill Moyers, *The Power of Myth*

[65] Karen Anderson, *A Short History of Myth*

[66] George Carey, *A Far Away Time and Place*

[67] George Carey, *A Far Away Time and Place*

[68] George Carey, *A Far Away Time and Place*

[69] George Carey, *A Far Away Time and Place*

[70] George Carey, *A Far Away Time and Place*

[71] Margot Adler, *Drawing Down the Moon*

[72] Margot Adler, *Drawing Down the Moon*

[73] According to string theory, the shape of the strings defines the quark and then atom, and scientists refer to the strings or building blocks of matter as "shape changers."

[74] George Carey, *Maryland Folklore* and *A Far Away Time and Place*

[75] George Carey, *A Far Away Time and Place*

[76] George Carey, *Maryland Folklore* and *A Far Away Time and Place*

[77] Alice Anne Parker, *Understanding Your Dreams*

[78] Thomas Flowers, *Shore Folklore*

[79] Margot Adler, *Drawing Down the Moon*

[80] George Carey, *A Far Away Time and Place, Lore of the Eastern Shore*

[81] Dickson Preston, *Trappe: The Story of an Old Fashioned Town*

[82] *The County Courthouses and Records of Maryland, Part One: The Courthouses*

[83] *The County Courthouses and Records of Maryland, Part One: The Courthouses*

[84] William Shakespeare, *Othello*

[85] Christopher Weeks, *Between the Nanticoke and the Choptank, An Architectural History of Dorchester County, Maryland*

[86] William Shakespeare, *Macbeth*

[87] James Ryder Randall, *Maryland, My Maryland*

[88] Thornton Wilder, *Our Town*

The future of publishing...today!

Apprentice House is the country's only campus-based, student-staffed book publishing company. Directed by professors and industry professionals, it is a nonprofit activity of the Communication Department at Loyola University in Maryland.

Using state-of-the-art technology and an experiential learning model of education, Apprentice House publishes books in untraditional ways. This dual responsibility as publishers and educators creates an unprecedented collaborative environment among faculty and students, while teaching tomorrow's editors, designers, and marketers.

Outside of class, progress on book projects is carried forth by the AH Book Publishing Club, a co-curricular campus organization supported by Loyola University's Office of Student Activities.

Student Project Team for *Dredging the Choptank*
  Nicole Trombly '09   Emily Rezin '12
  Regina Lyons '08   Lauren Crewell '10

Eclectic and provocative, Apprentice House titles intend to entertain as well as spark dialogue on a variety of topics. Financial contributions to sustain the press's work are welcomed. Contributions are tax deductible to the fullest extent allowed by the IRS.

To learn more about Apprentice House books or to obtain submission guidelines, please visit www.ApprenticeHouse.com.

Apprentice House
Communication Department
Loyola University in Maryland
4501 N. Charles Street
Baltimore, MD 21210
Ph: 410-617-5265 • Fax: 410-617-5040
info@apprenticehouse.com

www.ingramcontent.com/pod-product-compliance
Lightning Source LLC
Chambersburg PA
CBHW031231090426
42742CB00007B/152